# Intelligent Systems, Control and Automation: Science and Engineering

Volume 97

***Intelligent Systems, Control and Automation: Science and Engineering*** book series publishes books on scientific, engineering, and technological developments in this interesting field that borders on so many disciplines and has so many practical applications: human-like biomechanics, industrial robotics, mobile robotics, service and social robotics, humanoid robotics, mechatronics, intelligent control, industrial process control, power systems control, industrial and office automation, unmanned aviation systems, teleoperation systems, energy systems, transportation systems, driverless cars, human-robot interaction, computer and control engineering, but also computational intelligence, neural networks, fuzzy systems, genetic algorithms, neurofuzzy systems and control, nonlinear dynamics and control, and of course adaptive, complex and self-organizing systems. This wide range of topics, approaches, perspectives and applications is reflected in a large readership of researchers and practitioners in various fields, as well as graduate students who want to learn more on a given subject. The series has received an enthusiastic acceptance by the scientific and engineering community, and is continuously receiving an increasing number of high-quality proposals from both academia and industry. The current Series Editor is Kimon Valavanis, University of Denver, Colorado, USA. He is assisted by an Editorial Advisory Board who help to select the most interesting and cutting edge manuscripts for the series: Panos Antsaklis, University of Notre Dame, USA Stjepan Bogdan, University of Zagreb, Croatia Alexandre Brandao, UFV, Brazil Giorgio Guglieri, Politecnico di Torino, Italy Kostas Kyriakopoulos, National Technical University of Athens, Greece Rogelio Lozano, University of Technology of Compiegne, France Anibal Ollero, University of Seville, Spain Hai-Long Pei, South China University of Technology, China Tarek Sobh, University of Bridgeport, USA.

Springer and Professor Valavanis welcome book ideas from authors. Potential authors who wish to submit a book proposal should contact Thomas Ditzinger (thomas.ditzinger@springer.com)

Indexed by SCOPUS, Google Scholar and SpringerLink.

More information about this series at http://www.springer.com/series/6259

Xinwei Wang · Jie Liu · Haijun Peng

# Symplectic Pseudospectral Methods for Optimal Control

## Theory and Applications in Path Planning

Springer

Xinwei Wang
Department of Engineering Mechanics
Dalian University of Technology
Dalian, Liaoning, China

Jie Liu
War Research Institute
Academy of Military Sciences
Beijing, China

Haijun Peng
Department of Engineering Mechanics
Dalian University of Technology
Dalian, Liaoning, China

ISSN 2213-8986 ISSN 2213-8994 (electronic)
Intelligent Systems, Control and Automation: Science and Engineering
ISBN 978-981-15-3437-9 ISBN 978-981-15-3438-6 (eBook)
https://doi.org/10.1007/978-981-15-3438-6



This Springer imprint is published by the registered company Springer Nature Singapore Pte Ltd.
The registered company address is: 152 Beach Road, #21-01/04 Gateway East, Singapore 189721, Singapore

# Preface

Vehicles are equipment with capabilities to realize certain missions in various workspace such as land, sea, air, and space. Their structures usually exhibit complex dynamical characteristics and dynamical behaviors. Mechanical disciplines such as analytical dynamics, multi-body dynamics, and nonlinear dynamics give deep insights into motion laws of vehicles. Meanwhile, control disciplines provide feasible and powerful tools to regulate the motion of vehicles according to human will.

In the 1960s, Richard Bellman (academician of United States National Academy of Sciences) and Lev S. Pontryagin (academician of Academy of Sciences of former Soviet Union) had laid the mathematical foundation of optimal control theory. Optimal control theory greatly facilitated the progress of military science and equipment during the cold war, successively providing cutting edge technologies for the two countries for tens of years. Nowadays, the optimal control problems for vehicles encountered in engineering are becoming more and more complicated, which makes it impossible to implement analytical solutions by Bellman's dynamic programming and Pontryagin's maximum principle. Arthur E. Bryson (the academician of both United States National Academy of Sciences and American Academy of Arts and Sciences), who is considered as the father of modern optimal control theory, points out that "with powerful digital computer, numerical solutions can be found for realistic problems". It significantly prompts the development of computational optimal control.

The trajectory optimization of vehicles is a typical open-loop optimal control problem for nonlinear systems. It aims at finding optimal control inputs and corresponding trajectory to fulfill the mission, meanwhile minimizing a certain objective and satisfying various kinds of constraints. Direct methods and indirect methods are the two main categories of computational optimal control techniques. They both usually start from the difference discretization of dynamic equations while neglecting the inherent dynamical characteristics of optimal control problems. This leads to deficiencies in the stability and precision of computational optimal control techniques. Jerrold E. Marsden, the fellow of British Royal Society and the leading scholar in classical mechanics, says that "Algorithms could be developed with the natural dynamics built in, thereby yielding better convergence properties". In addition, Gene H. Golub, the founder of modern matrix computation, points out that "It is a basic tenet of numerical analysis that solution procedures should exploit structure whenever it is present." For optimal control problems, they have inherent Hamiltonian mathematical structures. Thus, the energy/momentum variation of the controlled system with respect to time can be seen as important *natural dynamics*, and the symplectic structure is a distinct *mathematical structure* in Hamiltonian systems.

State-space representation is the basis of optimal control theory and it can trace back to the system of Hamiltonian canonical equations. Hamiltonian systems, as the cornerstone in analytical dynamics, bridge the gap between optimal control and analytical mechanics. Based on Hamiltonian systems, Wanxie Zhong (the academician of Chinese Academy of Sciences) identified the simulation between structural mechanics and optimal control for the first time. And a series of computational optimal/robust control methods for linear systems have been proposed drawing ideas in computational mechanics.

The first and the third authors of the book, i.e., Xinwei Wang and Haijun Peng, were both taking Ph.D.s under the supervision of Prof. Zhong. In the last two decades, based on Hamiltonian systems and the symplectic theory, the research group has developed a series of symplectic algorithms to solve nonlinear optimal control problems with different complex features. In this book, we report our recent progress in symplectic numerical algorithms that incorporate pseudospectral methods to achieve better computational efficiency and accuracy. Additionally, Jie Liu, the second author of the book, has applied the developed symplectic pseudospectral method to solve trajectory optimization problems in various fields. Hence, some of his works are taken as examples in this book.

The publication of the book should appreciate those who helped us over the years. Our first gratitude goes to Prof. Zhong who was the pathfinder in our research direction. And many thanks go to other colleagues from the Dalian University of Technology, including Prof. Qiang Gao, Prof. Zhigang Wu, Prof. Shujun Tan, etc. Gratitude also goes to Prof. Wei Han from Naval Aviation University. Besides, many of our current and former students, such as Mingwu Li, Xin Jiang, Boyang Shi, also contributes much to the development of algorithms during the last decade. Finally, we gratefully acknowledge helpful suggestions and assistance by Jasmine Dou (Editor, Springer) and her staff.

Dalian, China
Beijing, China
Dalian, China
July 2020

Xinwei Wang
Jie Liu
Haijun Peng

# Contents

# Chapter 1
# Overview of This Book

## 1.1 Optimal Control

Optimal control is an important component of modern control theory [1, 2]. In the 1960s, the former Soviet Union scholar Lev Pontryagin proposed the Pontryagin's maximum principle [3], meanwhile, the American scholar Richard Bellman developed the dynamic programming method [4]. Such two techniques significantly enriched and improved the theory of optimal control, prompting the development of the analytical solution of optimal control problems. Humans were initially starting the exploration of the space at that time, techniques based on optimal control theory such as orbit design [5] and the Kalman filter [6] won great successes.

However, as optimal control problems encountered in engineering are getting more complicated (i.e., large-scale, complex constraints, time-delay, etc.), the analytical solution becomes impossible. Luckily, due to the rapid development of computational techniques and devices, it is available for engineers to implement numerical solutions. During the last 70 years, the optimal control theory has been widely applied to various fields such as aerospace engineering, chemical engineering, economics, communications, automobile engineering. Hence, computational optimal control techniques [7–9], as the core to implement optimal control theory in practice, have drawn more and more attention.

## 1.2 Pseudospectral Methods

Numerical methods for optimal control problems are generally divided into indirect methods and direct methods. Pseudospectral methods, as the most popular direct methods in the last two decades, have drawn much attention [10]. The collocation



X. Wang et al., *Symplectic Pseudospectral Methods for Optimal Control*, Intelligent Systems, Control and Automation: Science and Engineering 97,
https://doi.org/10.1007/978-981-15-3438-6_1

points in pseudospectral methods are generally orthogonal Gauss points, leading to a highly accurate approximation of the original problem. More precise solutions can be obtained by pseudospectral methods when compared to other direct schemes under the same scale of discretization. Additionally, pseudospectral methods have a faster convergent rate, i.e., they show exponential convergent rate for problems where solutions are smooth and well behaved. Though this excellent property loses for constrained problems, multi-interval strategies can be used for compensation, leading to local pseudospectral methods. The first successful practical application of pseudospectral methods is the zero propulsion attitude maneuver of Internal Space Station in 2007. The operation command generated by pseudospectral methods helps to save fuel that values about one million dollars [11].

## 1.3 The Property of Symplectic Conservation

Optimal control problems can be transformed into Hamiltonian systems by the Pontryagin's maximum principle or the variational principle. The most notable feature of Hamiltonian systems is the phase flow in Hamiltonian systems is a symplectic transformation. Numerical methods that are symplectic conservative can solve Hamiltonian systems efficiently [12]. Hence, computational optimal control methods owning the symplectic conservative property are much appealing.

The concept of symplectic conservation is first proposed in computational mechanics. Hence, many scholars draw on mature theories in computational mechanics to enrich the symplectic conservative application in computational optimal control. For example, Zhong et al., explored the simulation between computational structure mechanics and optimal control [13]. Marsden extended the variational integration methods in computational dynamics to computational optimal control and proposed the Discrete Mechanics and Optimal Control (DMOC) method based on the Lagrange-D'Alembert's principle in analytical mechanics [14, 15]. Recently, Peng et al. developed a series of symplectic methods based on the generating function method [16–19].

Computational techniques that own the property of symplectic conservation has three most notable merits. First, the Hamiltonian structure of the original system is conserved; Secondly, they have good stability for problems with a long time interval; Thirdly, they can precisely reflect the energy variation for mechanical systems. Hence, it is attractive to construct symplectic numerical methods for optimal control problems.

## 1.4 Motivation of the Book

For decades, pseudospectral methods are generally developed and improved under the framework of direct methods. However, as essentially a numerical approximation technique, the application of pseudospectral methods should not be limited to the construction of direct methods. Additionally, during the design of traditional numerical methods, one usually focuses on how to improve the numerical precision under given discretization scheme while the inherent mathematical structures of optimal control problems are neglected. Based on the above facts, this book will start from the inherent Hamiltonian mathematical structure of optimal control problems and develop a series of symplectic pseudospectral methods (SPMs) for problems with different features. The parametric variational principle and the multi-interval pseudospectral methods are used when constructing SPMs. Additionally, the SPM is taken as the core solver to construct symplectic pseudospectral model predictive controllers. Finally, some successful applications of SPMs in solving path planning problems are presented.

## 1.5 Scope of the Book

The book is constituted of three parts. Part I (this chapter, Chaps. 2, and 3) is the introductory material. Part II (Chaps. 4–7) provides a series of symplectic pseudospectral methods for nonlinear optimal control problems with various complicated factors. Part III (Chaps. 8–11) gives the application of symplectic pseudospectral methods in trajectory planning of various vehicles. The detailed content of each chapter is summarized as follows:

**Part I: Introductory materials**

This chapter gives an overview of this book, where the motivation of this book is emphasized.

Chapter 2 summarizes the numerical methods for nonlinear optimal control problems. Four kinds of computational techniques, i.e., indirect methods, direct methods, hybrid methods, and artificial intelligence-based methods are reviewed. And a brief comparison between indirect methods and direct methods is provided.

Chapter 3 first gives the mathematical formulations of three kinds of problems studied in this book. And mathematical foundations required when constructing SPMs, such as the Hamiltonian structure of optimal control problems, symplectic theory and pseudospectral methods are briefly presented.

**Part II: Symplectic pseudospectral methods for nonlinear optimal control problems**

Chapter 4 focuses on generally unconstrained nonlinear optimal control problems. An SPM for this kind of problem is developed based on the first kind of generating function. Besides, a mesh refinement technique based on the relative curvature of state variables is provided.

Chapter 5 focuses on problems with inequality constraints.

Chapter 6 focuses on time-delayed problems.

Chapter 7 presents the basic idea of model predictive control. And the SPM developed in Chap. 5 is taken as the core solver to construct the symplectic pseudospectral model predictive controller.

**Part III: Applications in trajectory planning and control**

Chapter 8 provides several applications in the optimal maneuver of the spacecraft.

Chapter 9 presents trajectory planning problems for unmanned ground systems with different configurations.

Chapter 10 presents an autonomous control framework of three-dimensional overhead cranes.

Chapter 11 focuses on trajectory planning issues for tractor-trailer systems.

## References

1. Aschepkov LT, Dolgy DV, Kim T, Agarwal RP (2016) Optimal control. Springer, Cham
2. Bowon Kim (2017) Optimal control applications for operations strategy. Springer, Cham
3. Pontryagin LS (1962) The mathematical theory of optimal processes (Trirogoff KN, trans). Interscience, New York
4. Bellman RE, Deryfus SE (1962) Applied dynamic programming. Princeton University Press, New Jersey
5. Moiseev NN, Lebedev VN (1966) Review paper on the research completed at the Computing Center of the Academy of Sciences of the USSR on the theory of optimal control functions of spacecraft. In: Aseltine JA (ed) Peaceful uses of automation in outer space. Springer, Boston
6. Kalman RE, Bucy RS (1961) New results in linear filtering and prediction theory. Trans ASME J Basic Eng 83:95–108
7. Mufti IH (1970) Computational methods in optimal control problems. Springer, Berlin
8. Bulirsch R, Kraft D (1994) Computational optimal control. Birkhäuser, Basel
9. Seywald H, Kumar RR (1997) Some recent developments in computational optimal control. In: Biegler LT, Coleman TF, Conn AR, Santosa FN (eds) Large-scale optimization with applications. The IMA volumes in mathematics and its applications, vol 93. Springer, New York
10. Ross IM, Karpenko M (2012) A review of pseudospectral optimal control: From theory to flight. Ann Rev Control 36(2):182–197
11. Kang W, Bedrossian N (2008) Pseudospectral optimal control theory makes debut flight, Saves NASA $1 M in Under Three Hours. SIAM News 40(7)
12. Arnold VI (1978) Mathematical methods of classical mechanics. Springer, New York
13. Zhong W (2004) Duality system in applied mechanics and optimal control. Springer, Boston
14. Junge O, Marsden JE, Ober-Blöbaum S (2005) Discrete mechanics and optimal control. IFAC Proc Volumes 38(1):538–543
15. Leyendecker S, Ober-Blöbaum S, Marsden JE et al (2010) Discrete mechanics and optimal control for constrained systems. Opt Control Appl Methods 31(6):505–528
16. Peng H, Gao Q, Wu Z et al (2013) Efficient sparse approach for solving receding-horizon control problems. J Guid Control Dyn 36(6):1864–1872
17. Peng H, Gao Q, Zhang H et al (2014) Parametric variational solution of linear-quadratic optimal control problems with control inequality constraints. Appl Math Mech (English Ed) 35(9):1079–1098
18. Peng H, Gao Q, Wu Z et al (2015) Symplectic algorithms with mesh refinement for a hypersensitive optimal control problem. Int J Comput Math 92(11):2273–2289
19. Peng H, Tan S, Gao Q et al (2017) Symplectic method based on generating function for receding horizon control of linear time-varying systems. Eur J Control 33:24–34

# Chapter 2
# Computational Techniques for Nonlinear Optimal Control

## 2.1 Introduction

There are already many excellent reviews on numerical techniques for nonlinear optimal control [1–5]. According to the implementation of numerical techniques, computational methods for optimal control problems can generally fall into two groups, i.e., direct methods and indirect methods. Besides, hybrid methods and artificial intelligence-based methods are also popular. In this chapter, the pros and cons of these methods are presented and researches that consider the property of symplectic conservation are reviewed.

## 2.2 Indirect Methods

Indirect methods transform an optimal control problem into a Hamiltonian two-point boundary value problem (TPBVP) of state and costate variables by the variational principle or the Pontryagin's maximum principle. The TPBVP may couple with other algebraic equations if complicated factors are involved in the optimal control problem, and it is called the first-order necessary conditions [6]. Then the numerical solver for TPBVPs is used to solve the problem.

The main advantage of indirect methods is twofold. On the one hand, numerical solutions obtained by the indirect methods are naturally local optimal since it is constructed based on the first-order necessary conditions. On the other hand, the solutions of costate variables are usually directly provided, which are the prerequisite to analyze the Hamiltonian structure of the original optimal control problem. As for the disadvantage of indirect methods, the most notable one is that one must derive the first-order necessary conditions of the problem according to its formulation,



X. Wang et al., *Symplectic Pseudospectral Methods for Optimal Control*, Intelligent Systems, Control and Automation: Science and Engineering 97,
https://doi.org/10.1007/978-981-15-3438-6_2

while this process would be extremely hard once complicated factors are involved. Fortunately, many mathematicians have done much fundamental work in this area, facilitating successive researchers to conduct deeper research under the framework of indirect methods. In [6], the first-order necessary conditions for problems with factors such as equality constraints, inequality constraints, interior point constraints, and integral constraints are provided. [7] focus on time-delayed optimal control problems. The first-order necessary conditions for fractional order optimal control problems are studied in [8, 9]. Another drawback of indirect methods is that high-quality initial guesses of state and costate variables are usually required. However, it is generally hard to provide a high-quality initial guess of costate variables since it is of no physical meaning.

For the transformed TPBVP, numerical methods such as shooting method [10–12], multiple shooting method [13], generating function method [14–17], and finite difference method [18–20], can be used. In shooting methods, initial guesses on costate variables must satisfy the transversality conditions. For problems with a short time interval, the shooting method is effective; however, for problems with a long time interval, it is prone to fail the convergence due to the numerical illness. It is seen that shooting methods are with small convergent radius and extremely sensitive to the initial guesses. Compared to shooting methods, the demand for initial guesses is not that high in the generating function method. However, plenty of series expansions are required in the generating method, one must turn to mathematical software such as MATLAB for help.

It should be noted that the Hamiltonian TPBVP structure naturally exists under the framework of indirect methods, which tends to facilitate the construction of symplectic methods. To realize the symplectic solving of linear-quadratic optimal control problems, Zhuk et al. design the symplectic Möbius integrator to solve Riccati equations in [21]. The generating function method mentioned above is also symplectic. Besides the initial development of Park et al., Pent et al. conduct some creative work in this area in recent years [22–25]. Focusing on the unconstrained nonlinear optimal control problem, they use the multiple interval mesh with regular Lagrange interpolation to transform the original problem into a system of nonlinear algebraic equations. Since the methods are constructed based on the least action principle and the finite element discretization, the core matrix is naturally sparse and symmetric, which is an advantage for solving large-scale problems. Later, Li et al. approximate state and costate variables by pseudospectral methods instead of the regular Lagrange interpolation [26]. Results suggest that discretization with pseudospectral methods lead to better computational efficiency and precision. In addition, focusing on problems with inequality constraints, Li et al. first transform the original nonlinear problem into a series of linear-quadratic problems by the quasilinearization technique. Then the transformed linear-quadratic problems are solved by symplectic algorithms in an iterative manner [27, 28]. The quasilinearization technique used in [27, 28] is actually a successive convexification technique. Due to its benefit, the initial guesses on costate variables are avoided, and the method is less sensitive to initial guesses.

## 2.3 Direct Methods

Direct methods transform a nonlinear optimal control problem into a finite-dimensional nonlinear programming (NLP) by discretization or parametrization [29]. Thus, NLP solvers can be used to solve the problem efficiently. The most notable advantage of direct methods is that they can treat the problem under a uniform NLP framework regardless of the characteristics of the problem. Optimal control problems met in engineering applications usually are subjected to complicated factors. Direct methods are popular in engineering due to this uniformity. In addition, direct methods usually have a larger convergent radius than indirect methods, which suggests they are less sensitive to initial guesses.

The disadvantages of direct methods mainly come down to three aspects. First, the scale of the resulted NLP grows faster than the number of variables, leading to the so-called "curse of dimensionality" [29]. Though computational devices nowadays have more and more computational power, this is still a bottleneck for large-scale problems. Secondly, constraints are usually slightly relaxed in direct method for numerical computation. Hence, original constraints may not be strictly satisfied in the obtained results. Thirdly, solutions obtained by direct methods usually do not satisfy the first-order necessary conditions for optimal control problems, which suggests that it even cannot guarantee the local optimality of the solutions. Information on costate variables is usually not directly available in direct methods, making it hard to analyze the Hamiltonian structure of the optimal control problem.

There are mainly two manners to implement the transformation to NLP by parametrization, i.e., function expansion [30, 31] and function interpolation. And the functional interpolation is commonly used. In methods based on function interpolation, control and/or state variables are approximated by a set of trial functions. The system equations and constraints are satisfied at a set of collation nodes within the time domain, and the cost functional is transformed as a function of state and control variables at the interpolation nodes. If the solution domain is divided into sub-intervals where state and controls are interpolated within the sub-interval, it leads to the local interpolation method. Conversely, if the solution domain is seen as a whole interval and the variables are interpolated within this big interval, it leads to the global interpolation method.

There are also some researches considering the symplectic conservation under the framework of direct methods. In [32], Bonnans discusses the symplectic conservation condition of using the Runge–Kutta difference scheme to discretize unconstrained nonlinear optimal control problems. The discrete mechanics and optimal control method (DMOC) mentioned in Sect. 1.3, is also a symplectic method, where the variational principle in discrete fashion is applied to achieve symplectic conservation [33, 34]. However, it should be noted that only symplectic conservation in system equations is considered. Hence, we say that the symplectic-conservation property in the DOMC method is not complete.

The book aims at developing symplectic methods that integrate the pseudospectral approximation. Hence, we will give a detailed introduction of pseudospectral methods for solving nonlinear optimal control problems in the rest of this section.

In classical pseudospectral methods, state and control variables are globally interpolated by orthogonal polynomials, such as Legendre polynomials and Chebyshev polynomials. System equations and constraints are imposed at collocation points based on Gauss integration points. The interpolation points and integration nodes are generally identical [35]. According to the selection of interpolation nodes, various pseudospectral methods such as Chebyshev-Gauss-Lobatto (CGL) pseudospectral method [36], Legendre-Gauss (LG) pseudospectral method [37, 38], Legendre-Gauss-Radau (LGR) pseudospectral method [39–41], and Legendre-Gauss-Lobatto (LGL) pseudospectral method [42–44], are developed. In [45], Garg et al. provide a uniform framework for LG, LGR, and LGL pseudospectral methods to solve nonlinear optimal control problems.

The first-order necessary conditions are not satisfied in solutions obtained by most direct methods. However, it has been proven that the Karush-Kuhn-Tucker (KKT) conditions of the NLP obtained by pseudospectral methods are equivalent to the first-order necessary conditions. Besides, the costate mapping theory for Legendre pseudospectral methods and Chebyshev pseudospectral methods are developed [39, 46, 47], making it possible to obtain the information on costate variables. These researches bridge the gap between pseudospectral methods and indirect methods.

For problems whose solutions are smooth and well behaved, pseudospectral methods exhibit an exponential convergent rate. However, for problems with constraints, the exponential convergent rate is lost. Hence, some researches first divide the solution domain into sub-intervals and then apply pseudospectral approximation within each sub-interval, leading to the *local pseudospectral methods*. Compared to local pseudospectral methods, classical pseudospectral methods can be seen as *global pseudospectral methods*. In local pseudospectral methods (corresponding to the $h$ method in finite element analysis), the solution domain is divided into sub-intervals, and the degree of the pseudospectral approximation in each sub-interval is usually low. Thus, convergence is achieved by increasing the number of sub-intervals. As for global pseudospectral methods (corresponding to the $p$ method in finite element analysis), convergence is achieved by increasing the approximation degree.

Either the $h$ method or the $p$ method has its drawbacks. In $h$ methods, it may generate too many sub-intervals. And in $p$ methods, it may result in an unreasonably high approximation degree. Hence, some researches combine the characteristics of both the $h$ method and the $p$ method to construct $hp$ methods for solving nonlinear optimal control problems [40, 47, 48]. In an $hp$ method, both the number of sub-intervals and the approximation degree in each sub-interval are tunable. It should be noted that the manual mesh refinement could be blind. Hence, it is necessary to develop adaptive mesh refinement techniques to make the refinement process reasonable and efficient. In [38, 41, 49], mesh refinement techniques based on the smoothness of solutions are developed, dramatically accelerating the convergence process.

## 2.4 Hybrid Methods

There are typically two ways to construct hybrid methods for solving nonlinear optimal control problems.

Firstly, considering that direct methods usually have a larger convergent radius than indirect methods, one can first obtain a solution by direct methods and then take it to initialize the solving process of indirect methods. Thus, the first-order necessary conditions are satisfied in the final solutions. Additionally, indirect methods have inherent drawbacks to solve singular optimal control problems. Focusing on the fuel optimal problem, Li et al. [50] first solve the relatively easier energy optimal control problem by the pseudospectral method. Then, by taking such solutions as initial guesses and incorporating with the homotopy technique, the fuel optimal control problem is solved by indirect methods. In [51], they generate initial guess of costate variables of high precision by the costate mapping theory and the quasilinearization techniques are used to further facilitate the convergence process.

The second manner focuses on solving the resultant TPBVP from the view of optimization. In [52], Enright et al. first derive the terminal boundary conditions by the Pontryagin's maximum principle. An optimization problem where the quadratic of terminal boundary conditions is taken as the objective to be minimized is formulated and then solved by the collocation method.

It is seen that the main purpose of such two manners is to overcome the drawback of convergence of indirect methods. Hybrid methods break the boundaries between indirect and direct methods, facilitating researches to construct the solving strategy that takes advantages of both the two categories of methods. Both of the two manners mentioned above focus on how to efficiently solve the resulted TPBVP robustly and efficiently. Hence, hybrid methods are generally thought much alike to indirect methods essentially.

## 2.5 Artificial Intelligence-Based Methods

With the rapid development of artificial intelligence (AI), more and more disciplines have introduced AI techniques to solve problems that are hard to be solved by traditional methods. The main research directions solving optimal control problems by AI fall into the following two categories.

On the one hand, some researches use mature evolutionary algorithms (e.g., genetic algorithm, particle swarm optimization, ant colony algorithm, etc.) or their improved variants to solve optimal control problems. Both single-objective optimization [53–56] and multi-objective optimization [57–59], are successfully implemented to solve complex problems which are difficult to solve by traditional methods. Under this framework, trajectory optimization problems of mechanical systems, especially spacecraft, have made much progress [60–62]. Nowadays, the Global Trajectory Optimization Problem (GTOP) database has collected many cases, and they are often taken as benchmarks for novel evolutionary algorithms [63]. Evolutionary algorithms theoretically can avoid falling into local optimum due to their ability for global optimization. In addition, gradient information is not required in evolutionary algorithms and there are many mature software packages available, which allows researchers without much background in control theory to solve complex nonlinear optimal control problems easily. However, computational efficiency is the bottleneck that restrains the application for evolutionary algorithms to solve problems of large-scale. Hence, it seems to be a wise idea to adopt a hierarchical strategy to solve complex optimal control problems. First, we discretize the problem at a relatively small scale and obtain a coarse solution by evolutionary algorithms. Then, we take such a coarse solution as initial guesses and solve the problem by indirect or direct methods. The high-quality initial guesses prompt the convergence rate. Moreover, the global optimum is guaranteed to a great extent.

On the other hand, machine learning techniques (such as artificial neural network, decision tree, random forest, etc.) have been more and more used in solving optimal control problems [64–68]. For most machine learning-based methods, a large amount of samples and/or a long time are required for training. However, such methods provide an efficient manner to solve problems where the system equations are not easy to obtained [69]. Under this framework, plenty of trials on guidance and control for spacecraft have been made. Machine learning techniques have shown their unique advantages in various engineering fields. And we deeply believe that it will greatly prompt the numerical solving for optimal control problems in the near future.

**Table 2.1** Comparison of indirect methods and direct methods from various aspects

| Property | Direct methods | Indirect methods |
|---|---|---|
| Mathematical formulation | Easy | Hard |
| Sensitive to initial guesses | Low | High |
| Handling constraints | Easy | Hard |
| Numerical precision | High | Very high |
| Engineering applications | Many | Less (almost limited to orbit design) |
| Mature software packages | Many | Very few |

## 2.6 A Brief Conclusion

Direct methods and indirect methods, as the two main numerical techniques to solve nonlinear optimal control problems, have their pros and cons. In Table 2.1, we compare such two kinds of numerical methods from various aspects.

Finally, it should be noted that though various numerical techniques for solving optimal control problems are proposed and continuously being improved, there is not a certain method that can perfectively solve optimal control problems of any kind. Hence, during the practical solving process, one should carefully analyze the feature of a problem and choose proper solving strategies.

## References


1. Polak E (1973) An historical survey of computational methods in optimal control. SIAM Rev 15(2):553–584
2. Stryk OV, Bulirsch R (1992) Direct and Indirect Methods for Trajectory Optimization. Ann Oper Res 37(1):357–373
3. Betts JT (1998) Survey of numerical methods for trajectory optimization. J Guid Control Dyn 21(2):193–207
4. Rao AV (2009) A survey of numerical methods for optimal control. In: AAS/AIAA astrodynamics specialist conference
5. Conway BA (2012) A survey of methods available for the numerical optimization of continuous dynamic systems. J Optim Theory Appl 152(2):271–306
6. Bryson AE, Ho YC (1975) Applied optimal control. Hemisphere/Wiley, New York
7. Kharatishvili GL (1961) The maximum principle in the theory of optimal process with time-lags. Dokl Akad Nauk SSSR 136:39–42
8. Agrawal OP (2004) A general formulation and solution scheme for fractional optimal control problems. Nonlinear Dyn 38(1–4):323–337
9. Jelicic ZD, Petrovacki N (2009) Optimality conditions and a solution scheme for fractional optimal control problems. Struct Multi Optim 38(6):571–581
10. Lastman GJ (1978) A shooting method for solving two-point boundary-value problems arising from non-singular bang-bang optimal control problems. Int J Control 27(4):513–524
11. Holsapple R, Venkataraman R, Doman DB (2003) A modified simple shooting method for solving two-point boundary-value problems. In: IEEE aerospace conference
12. Fotouhic R, Szyszkowski W (2002) A numerical approach for time-optimal control of double arms robot. In: IEEE conference on control applications
13. Fabien BC (1996) Numerical solution of constrained optimal control problems with parameters. Appl Math Comput 80(1):43–62
14. Park C, Scheeres DJ (2005) Extended applications of generating functions to optimal feedback control problems. In: American control conference
15. Park C, Scheeres DJ (2006) Determination of optimal feedback terminal controllers for general boundary conditions using generating functions. Automatica 42(5):869–875
16. Park C, Scheeres D, Guibout V (2013) Solving optimal continuous thrust rendezvous problems with generating functions. J Guid Control Dyn 37(29):396–401
17. Lee K, Park C, Park SY (2013) Performance analysis of generating function approach for optimal reconfiguration of formation flying. J Astron Space Sci 30(1):17–24
18. Marzban HR, Hoseini SM (2013) A composite Chebyshev finite difference method for nonlinear optimal control problems. Commun Nonlinear Sci Numer Simul 18(6):1347–1361

19. Nikoobin A, Moradi M (2015) Indirect solution of optimal control problems with state variable inequality constraints: finite difference approximation. Robotica 35(1):50–72
20. Jajarmi A, Hajipour M (2017) An efficient finite difference method for the time-delay optimal control problems with time-varying delay. Asian J Control 19(2):554–563
21. Zhuk S, Frank J (2014) Symplectic Möbius integrators for LQ optimal control problems. In: IEEE decision and control conference
22. Peng H, Gao Q, Wu Z et al (2013) Efficient sparse approach for solving receding-horizon control problems. J Guid Control Dyn 36(6):1864–1872
23. Peng H, Gao Q, Zhang H et al (2014) Parametric variational solution of linear-quadratic optimal control problems with control inequality constraints. Appl Math Mech (Engl Ed) 35(9):1079–1098
24. Peng H, Gao Q, Wu Z et al (2015) Symplectic algorithms with mesh refinement for a hypersensitive optimal control problem. Int J Comput Math 92(11):2273–2289
25. Peng H, Tan S, Gao Q et al (2017) Symplectic method based on generating function for receding horizon control of linear time-varying systems. Eur J Control 33:24–34
26. Li M, Peng H, Wu Z (2015) Symplectic irregular interpolation algorithms for optimal control problems. Int J Comput Methods 12(06):1550040
27. Li M, Peng H, Zhong W (2015) A symplectic sequence iteration approach for nonlinear optimal control problems with state-control constraints. J Franklin Inst 352(6):2381–2406
28. Li M, Peng H (2016) Solutions of nonlinear constrained optimal control problems using quasilinearization and variational pseudospectral methods. ISA Trans 62:177–192
29. Betts JT (2001) Practical methods for optimal control using nonlinear programming. SIAM, Philadelphia
30. Vlassenbroeck J (1988) A chebyshev polynomial method for optimal-control with state constraints. Automatica 24(4):499–506
31. Marzban HR, Razzaghi M (2010) Rationalized Haar approach for nonlinear constrained optimal control problems. Appl Math Model 34(1):174–183
32. Bonnans JF, Laurent-Varin J (2006) Computation of order conditions for symplectic partitioned Runge-Kutta schemes with application to optimal control. Numer Math 103(1):1–10
33. Junge O, Marsden JE, Ober-Blöbaum S (2005) Discrete mechanics and optimal control. IFAC Proc Volumes 38(1):538–543
34. Leyendecker S, Ober-Blöbaum S, Marsden JE et al (2010) Discrete mechanics and optimal control for constrained systems. Opt Control Appl Methods 31(6):505–528
35. Shen J, Tang T, Wang L (2011) Spectral methods: algorithms, analysis and applications. Springer, Berlin
36. Fahroo F, Ross IM (2000) Direct trajectory optimization by a Chebyshev pseudospectral method. In: American control conference
37. Thorvaldsen TP, Huntington GT, Benson DA (2006) Direct Trajectory Optimization and Costate Estimation via an Orthogonal Collocation Method. J Guid Control Dyn 29(6):1435–1439
38. Darby CL, Hager WW, Rao AV (2011) An hp-adaptive pseudospectral method for solving optimal control problems. Opt Control Appl Methods 32(4):476–502
39. Garg D, Patterson M, Francolin C et al (2011) Direct trajectory optimization and costate estimation of finite-horizon and infinite-horizon optimal control problems via a Radau pseudospectral method. Comput Optim Appl 49(2):335–358
40. Darby CL (2011) Hp-pseudospectral method for solving continuous-time nonlinear optimal control problems [thesis]. Department of Mechanical and Aerospace Engineering, University of Florida
41. Patterson MA, Hager WW, Rao AV (2015) A ph mesh refinement method for optimal control. Opt Control Appl Methods 36:398–421
42. Elnagar GN, Kazemi MA, Razzaghi M (1995) The pseudospectral Legendre method for discretizing optimal control problems. IEEE Trans Autom Control 40(10):1793–1796
43. Elnagar GN, Kazemi MA (1998) Pseudospectral Legendre-based optimal computation of nonlinear constrained variational problems. J Comput Appl Math 88(2):363–375

44. Ross IM, Fahroo F (2004) Pseudospectral methods for optimal motion planning of differentially flat systems. IEEE Trans Autom Control 49(8):1410–1413
45. Garg D, Patterson MA, Hager WW et al (2010) A unified framework for the numerical solution of optimal control problems using pseudospectral methods. Automatica 46(11):1843–1851
46. Qi G, Ross IM, Fahroo F (2010) Costate Computation by a Chebyshev pseudospectral method. J Guid Control Dyn 33(2):623–628
47. Darby CL, Garg D, Rao AV (2015) Costate estimation using multiple-interval pseudospectral methods. J Spacecraft Rockets 48(5):856–866
48. Wei J, Tang X, Jie Y (2016) Costate estimation for a multiple-interval pseudospectral method using collocation at the flipped legendre-Gauss-Radau points. IEEE/CAA J Automatica Sinica 99:1–15
49. Patterson MA, Rao AV (2010) GPOPS-II: a MATLAB software for solving multiple-phase optimal control problems using hp-adaptive Gaussian quadrature collocation methods and sparse nonlinear programming. ACM Trans Math Softw 41(1):1–37
50. Guo T, Jiang F, Li J (2012) Homotopic approach and pseudospectral method applied jointly to low thrust trajectory optimization. Acta Astronaut
51. Tang G, Jiang F, Li J (2018) Fuel-optimal low-thrust trajectory optimization using indirect method and successive convex programming. IEEE Trans Aerosp Electron Syst 54(4):2053–2066
52. Enright PJ, Conway BA (1991) Optimal finite-thrust spacecraft trajectories using collocation and nonlinear programming. J Guidance Control Dyn 14(5):981–985
53. Radice G, Olmo G (2006) Ant colony algorithms for two impluse interplanetary trajectory optimization. J Guidance Control Dyn 29(6):1440–1444
54. Sentinella MR, Casalino L (2009) Hybrid evolutionary algorithm for the optimization of interplanetary trajectories. J Spacecraft Rockets 46(2):365–372
55. Englander JA, Conway BA, Williams T (2012) Automated mission planning via evolutionary algorithms. J Guidance Control Dyn 35(6):1878–1887
56. Pontani M, Conway BA (2013) Optimal finite-thrust rendezvous trajectories found via particle swarm algorithm. J Spacecraft Rockets 50(6):1222–1234
57. Deb K, Padhye N, Neema G (2007) Interplanetary trajectory optimization with swing-bys using evolutionary multi-objective optimization. In: ACM conference companion on genetic & evolutionary computation
58. Schütze O, Vasile M, Junge O et al (2009) Designing optimal low-thrust gravity-assist trajectories using space pruning and a multi-objective approach. Eng Opt 41(2):155–181
59. Izzo D, Hennes D, Riccardi A (2014) Constraint handling and multi-objective methods for the evolution of interplanetary trajectories. J Guidance Control Dyn
60. Girimonte D, Izzo D (2007) Artificial intelligence for space applications. In: Intelligent computing everywhere. Springer, Berlin, pp 235–253
61. Izzo D, Märtens M, Pan B (2018) A survey on artificial intelligence trends in spacecraft guidance dynamics and control. arXiv:1812.02948
62. Izzo D, Sprague C, Tailor D (2018) Machine learning and evolutionary techniques in interplanetary trajectory design. arXiv:1802.00180
63. Vinkó T, Izzo D (2008) Global optimisation heuristics and test problems for preliminary spacecraft trajectory design. European Space Agency, Advanced Concepts Team, ACT Technical Report, id: GOHTPPSTD
64. Sánchezsánchez C, Izzo D (2016) Real-time optimal control via deep neural networks: study on landing problems. J Guidance Control Dyn 41(3):1–14
65. Furfaro R, Bloise I, Orlandelli M, et al (2018) Deep learning for autonomous lunar landing. In: AAS/AIAA astrodynamics specialist conference
66. Shang H, Wu X, Qiao D et al (2018) Parameter estimation for optimal asteroid transfer trajectories using supervised machine learning. Aerosp Sci Technol 79:570–579
67. Chu X, Alfriend KT, Zhang J, et al (2018) Q-learning algorithm for pathplanning to maneuver through a satellite cluster. In: 2018 AAS/AIAA astrodynamics specialist conference, 2018, AAS 18–268

68. Izzo D, Tailor D, Vasileiou (2018) On the stability analysis of optimal state feedbacks as represented by deep neural models. arXiv: 1812.02532
69. Wei Q, Song R, Li B (2017) Self-learning optimal control of nonlinear systems, adaptive dynamic programming approach. Science Press & Springer, Beijing

# Chapter 3
# Mathematical Foundation

## 3.1 Introduction

The book focuses on the symplectic pseudospectral methods to solve nonlinear optimal control problems. And this section provides necessary mathematical foundations to understand how we construct symplectic pseudospectral methods, including (i) basic concepts of optimal control problems, (ii) mathematical foundations of symplectic methods, and (iii) implementations of pseudospectral methods.

This section is arranged as follows. In Sect. 3.2, we give the general mathematical formulations of the three kinds of nonlinear optimal control problems studied in this book. In Sect. 3.3, the Hamiltonian structure of optimal control problems is derived. The mathematical foundations of symplectic methods and pseudospectral methods are provided in Sects. 3.4 and 3.5, respectively.

## 3.2 Three Kinds of Nonlinear Optimal Control Problems Studied in This Book

In Part II of this book, we focus on numerical methods to solve nonlinear optimal control problems with various complicated factors, such as constraints and time-delays. In this section, the general formulations of such problems are given in advance. In addition, we explain what simplifications are made to facilitate the derivations in Part II.



X. Wang et al., *Symplectic Pseudospectral Methods for Optimal Control*, Intelligent Systems, Control and Automation: Science and Engineering 97,
https://doi.org/10.1007/978-981-15-3438-6_3

### 3.2.1 General Unconstrained Nonlinear Optimal Control Problems

Consider the following optimal control problem:

Find optimal control input $\mathbf{u}^*(t) \in \mathbb{C}([t_s, t_f], \mathbb{R}^p)$ and corresponding $\mathbf{x}^*(t) \in \mathbb{C}([t_s, t_f], \mathbb{R}^d)$ that minimize the following cost functional

$$J = \Psi(\mathbf{x}(t_f)) + \int_{t_s}^{t_f} \Phi(\mathbf{x}(t), \mathbf{u}(t), t)\mathrm{d}t \tag{3.1}$$

meanwhile, satisfying the system equation

$$\dot{\mathbf{x}} = \mathbf{f}(\mathbf{x}(t), \mathbf{u}(t), t) \tag{3.2}$$

and the initial conditions

$$\mathbf{x}(t_s) = \mathbf{x}_s \tag{3.3}$$

where $t_s$ and $t_f$ are the initial and the terminal instant of the control problem, and both the variables are considered fixed in this book; $\mathbf{x}(t) \in \mathbb{C}([t_s, t_f], \mathbb{R}^d)$ is the state variable, $\mathbf{x}_s \in \mathbb{R}^d$ represents the initial conditions of the state; $\mathbf{u}(t) \in \mathbb{C}([t_s, t_f], \mathbb{R}^p)$ is the control variable; $\mathbf{f} : \mathbb{R}^d \times \mathbb{R}^p \times \mathbb{R} \to \mathbb{R}^d$ is a function of the state, control, and time variables that used to describe the system equation; $\Phi : \mathbb{R}^d \times \mathbb{R}^p \times \mathbb{R} \to \mathbb{R}$ and $\Psi : \mathbb{R}^d \to \mathbb{R}$ are used to describe the process and terminal cost in the cost functional, respectively.

If at least one of the cost functional (3.1) or the system Eq. (3.2) is nonlinear, the problem turns out to be an unconstrained nonlinear optimal control problem. And only the non-singular problem will be discussed in Chap. 4.

### 3.2.2 Nonlinear Optimal Control Problems with Inequality Constraints

Consider the following optimal control problem:

Find optimal control input $\mathbf{u}^*(t) \in \mathbb{C}([t_s, t_f], \mathbb{R}^p)$ and corresponding $\mathbf{x}^*(t) \in \mathbb{C}([t_s, t_f], \mathbb{R}^d)$ that minimize the following cost functional

$$J = \Psi(\mathbf{x}(t_f)) + \int_{t_s}^{t_f} \Phi(\mathbf{x}(t), \mathbf{u}(t), t)\mathrm{d}t \tag{3.4}$$

meanwhile, satisfying the system equation

$$\dot{\mathbf{x}} = \mathbf{f}(\mathbf{x}(t), \mathbf{u}(t), t) \tag{3.5}$$

the inequality constraints

$$\mathbf{h}(\mathbf{x}, \mathbf{u}, t) \leq \mathbf{0} \tag{3.6}$$

and the initial conditions

$$\mathbf{x}(t_s) = \mathbf{x}_s \tag{3.7}$$

where the terminal time $t_f$ can be either fixed or free; $\mathbf{h} : \mathbb{R}^d \times \mathbb{R}^p \times \mathbb{R} \to \mathbb{R}^q$ is a $q$-dimensional vector that describes the inequality constraints. Any element of $\mathbf{h}$ can be in a pure-state, a pure-control or a mixed state-control form. Other variables involved have that similar meaning as those in Sect. 3.2.1.

If at least one of the cost functional (3.4), the system Eq. (3.5) or the inequality constraints is nonlinear, the problem becomes nonlinear. This problem is discussed in Chap. 5, where the terminal cost term is eliminated to facilitate the derivation.

### *3.2.3 Nonlinear Time-Delayed Optimal Control Problems*

Consider the following optimal control problem:

Find the optimal control input $\mathbf{u}^*(t) \in \mathbb{C}([t_s, t_f], \mathbb{R}^p)$ and corresponding $\mathbf{x}^*(t) \in \mathbb{C}([t_s, t_f], \mathbb{R}^d)$ that minimize the following cost functional

$$J = \Psi(\mathbf{x}(t_f)) + \int_{t_s}^{t_f} \Phi(\mathbf{x}(t), \mathbf{u}(t), t)\mathrm{d}t \tag{3.8}$$

meanwhile, satisfying the delayed system equation

$$\dot{\mathbf{x}} = \mathbf{f}(\mathbf{x}(t), \mathbf{x}(t - \tau_1), \ldots, \mathbf{x}(t - \tau_n), \mathbf{u}(t), \mathbf{u}(t - \upsilon_1), \ldots, \mathbf{u}(t - \upsilon_m), t) \tag{3.9}$$

and the initial conditions

$$\mathbf{x}(t) = \mathbf{x}_s(t), \quad \text{for } t_s - \tau_n \leq t \leq t_s \tag{3.10}$$

$$\mathbf{u}(t) = \mathbf{u}_s(t), \quad \text{for } t_s - \upsilon_m \leq t \leq t_s \tag{3.11}$$

where $\tau_i (i = 1, 2, \ldots, n)$ are delays in state variable which satisfy $\tau_1 < \tau_2 < \ldots < \tau_n$; $\upsilon_i (i = 1, 2, \ldots, m)$ are delays in control variables which satisfy $\upsilon_1 < \upsilon_2 < \ldots < \upsilon_m$; $\mathbf{x}_s(t) \in \mathbb{C}\big([t_s - \tau_n, t_s], \mathbb{R}^d\big)$ represents the initial conditions of states before $t_s$; $u_s(t) \in \mathbb{C}([t_s - \upsilon_m, t_s], \mathbb{R}^p)$ represents the initial conditions of controls before $t_s$. Other variables involved have that similar meaning as those in Sect. 3.2.1.

If at least one of the cost functional (3.8) or the system Eq. (3.9) is nonlinear, it results in a nonlinear problem. Under the framework of indirect methods, it's pretty complex to deal with control delay and multiple delays. Hence, the problem with quadratic cost functional and only one state delay is considered in Chap. 6, to simplify the derivation.

## 3.3 The Hamiltonian Structure of Optimal Control Problems

Taking the general unconstrained nonlinear optimal control problem in Sect. 3.2.1, as an example, we discuss the inherent Hamiltonian structure of optimal control problems in this section.

For the problem in Sect. 3.2.1, we first introduce the costate variable $\boldsymbol{\lambda}(t) \in \mathbb{C}\big([t_s, t_f], \mathbb{R}^d\big)$ to construct the following augmented cost functional $J_A$

$$J_A = \Psi\big(\mathbf{x}(t_f)\big) + \int_{t_s}^{t_f} \Phi(\mathbf{x}(t), \mathbf{u}(t), t) + \boldsymbol{\lambda}^{\mathrm{T}}(t)(\mathbf{f}(\mathbf{x}(t), \mathbf{u}(t), t) - \dot{\mathbf{x}}(t))\mathrm{d}t \tag{3.12}$$

Considering the first-order variation of Eq. (3.12), one has

$$\begin{aligned} \delta J_A = {} & \delta\mathbf{x}^{\mathrm{T}}(t_f)\left(\frac{\partial\Psi(\mathbf{x}(t_f), t_f)}{\partial\mathbf{x}(t_f)} - \boldsymbol{\lambda}(t_f)\right) + \delta\mathbf{x}^{\mathrm{T}}(t_s)\boldsymbol{\lambda}(t_s) \\ & + \int_{t_s}^{t_f}\left[-\delta\boldsymbol{\lambda}^{\mathrm{T}}(\dot{\mathbf{x}} - \mathbf{f}) + \delta\mathbf{x}^{\mathrm{T}}\left(\dot{\boldsymbol{\lambda}} + \frac{\partial\Phi}{\partial\mathbf{x}} + \boldsymbol{\lambda}^{\mathrm{T}}\frac{\partial\mathbf{f}}{\partial\mathbf{x}}\right) + \delta\mathbf{u}^{\mathrm{T}}\left(\frac{\partial\Phi}{\partial\mathbf{u}} + \boldsymbol{\lambda}^{\mathrm{T}}\frac{\partial\mathbf{f}}{\partial\mathbf{u}}\right)\right]\mathrm{d}t = 0 \end{aligned} \tag{3.13}$$

Thus, the first-order necessary conditions for the problem are obtained as follows:

$$\dot{\mathbf{x}} = \mathbf{f}(\mathbf{x}, \mathbf{u}, t) \tag{3.14}$$

$$\dot{\boldsymbol{\lambda}} = -\frac{\partial\Phi}{\partial\mathbf{x}} - \boldsymbol{\lambda}^{\mathrm{T}}\frac{\partial\mathbf{f}}{\partial\mathbf{x}} \tag{3.15}$$

$$\frac{\partial\Phi}{\partial\mathbf{u}} + \boldsymbol{\lambda}^{\mathrm{T}}\frac{\partial\mathbf{f}}{\partial\mathbf{u}} = \mathbf{0} \tag{3.16}$$

One can define the Hamiltonian of the problem as

$$\bar{H}(\mathbf{x}(t), \boldsymbol{\lambda}(t), \mathbf{u}(t), t) = \Phi(\mathbf{x}(t), \mathbf{u}(t), t) + \boldsymbol{\lambda}^{\mathrm{T}}(t)\mathbf{f}(\mathbf{x}(t), \mathbf{u}(t), t) \tag{3.17}$$

Thus, the first-order necessary conditions in Eqs. (3.14)–(3.16), can be expressed as

$$\dot{\mathbf{x}} = \frac{\partial \bar{H}(\mathbf{x}(t), \boldsymbol{\lambda}(t), \mathbf{u}(t), t)}{\partial \boldsymbol{\lambda}} \tag{3.18}$$

$$\dot{\lambda} = -\frac{\partial \bar{H}(\mathbf{x}(t), \boldsymbol{\lambda}(t), \mathbf{u}(t), t)}{\partial \mathbf{x}} \tag{3.19}$$

$$\frac{\partial \bar{H}}{\partial \mathbf{u}} = \mathbf{0} \tag{3.20}$$

According to Eq. (3.13), the terminal boundary conditions of the problem can be derived. For the problem with fixed terminal states, one has

$$\mathbf{x}(t_s) = \mathbf{x}_s,\ \mathbf{x}(t_f) = \mathbf{x}_g \tag{3.21}$$

where $\mathbf{x}_g$ is the prescribed terminal state. And for the problem with free terminal states, one has

$$\mathbf{x}(t_s) = \mathbf{x}_s,\ \boldsymbol{\lambda}(t_f) = \frac{\partial \Psi(\mathbf{x}(t_f), t_f)}{\partial \mathbf{x}(t_f)} \tag{3.22}$$

It is assumed that one can express the control variable $\mathbf{u}$ as the function of state and costate variables by solving Eq. (3.20), i.e., $\mathbf{u} = \mathbf{u}(\mathbf{x}, \boldsymbol{\lambda}, t)$. Thus, one can substitute the expression into Eq. (3.17), and the Hamiltonian can then be written as the functional of state and costate variables as $H(\mathbf{x}(t), \boldsymbol{\lambda}(t), t)$. In addition, the necessary conditions in Eqs. (3.18)–(3.19), can be written as the following Hamiltonian TPBVP

$$\dot{\mathbf{x}}(t) = \frac{\partial H(\mathbf{x}(t), \boldsymbol{\lambda}(t), t)}{\partial \boldsymbol{\lambda}(t)} \tag{3.23}$$

$$\dot{\lambda}(t) = -\frac{\partial H(\mathbf{x}(t), \boldsymbol{\lambda}(t), t)}{\partial \mathbf{x}(t)} \tag{3.24}$$

***Remark 3.1*** In the above derivations, we assume that the control variable can be expressed as a function of state and control variables by solving $\partial \bar{H} / \partial \mathbf{u} = \mathbf{0}$. Actually, it is requested that the second-order partial derivative of the Hamiltonian with respect to the control variable (i.e., $\bar{H}_{\mathbf{uu}}$) is full-rank [1]. Once this requirement is not satisfied, it will result in a singular optimal control problem. It is generally hard to solve a singular optimal control problem under the framework of indirect methods, and the homotopy technique is usually incorporated to facilitate the solving procedure [2]. Due to the complexity of singular optimal control problems, they are not considered in this book. □

It is seen that an optimal control problem owns inherent Hamiltonian structures where Eqs. (3.23)–(3.24) are called the Hamiltonian canonical equations. To simplify the construction of symplectic pseudospectral methods in the following chapters, the mathematical foundations of symplectic methods and pseudospectral methods are introduced in Sects. 3.4 and 3.5, respectively.

## 3.4 Mathematical Foundations of Symplectic Methods

This section introduces the mathematical foundations of symplectic methods involved in the derivation of the following chapters. In Sect. 3.4.1, the Hamiltonian dynamical system and the concept of symplectic conservation is provided. Then in Sect. 3.4.2, the concept of action and the least action principle is introduced. In Sect. 3.4.3, four kinds of generating functions and their applications in constructing symplectic methods for solving optimal control problems are provided.

It is noted that the contents in Sects. 3.4.2 and 3.4.3 are discussed under the general unconstrained nonlinear optimal control problems as provided in Sect. 3.2.1. For problems with complex factors such as inequality constrains and time-delays, their optimality conditions vary from each other. However, the core idea during the derivation is similar.

### 3.4.1 *Hamiltonian Dynamical Systems and the Property of Symplectic Conservation*

Similar to the first-order necessary condition with respect to state and control variables as given in Sect. 3.3, we define the following Hamiltonian canonical equations

$$\begin{bmatrix} \dot{\mathbf{q}}(t) \\ \dot{\mathbf{p}}(t) \end{bmatrix} = \begin{bmatrix} \frac{\partial H(\mathbf{q}(t),\mathbf{p}(t),t)}{\partial \mathbf{p}(t)} \\ -\frac{\partial H(\mathbf{q}(t),\mathbf{p}(t),t)}{\partial \mathbf{q}(t)} \end{bmatrix} \tag{3.25}$$

where $\mathbf{q}(t) \in \mathbb{R}^n$ and $\mathbf{p}(t) \in \mathbb{R}^n$ are called the generalized displacement and the generalized momentum, respectively, and $H(\mathbf{q}(t), \mathbf{p}(t), t)$ is the Hamiltonian. Once a dynamical system can be described by the canonical equations in Eq. (3.25), it is called a Hamiltonian system. The Hamiltonian canonical equations (3.25), utilize the dual variables, i.e., $\mathbf{q}(t)$ and $\mathbf{p}(t)$, to describe the behavior of a dynamical system. When defining the generalized state variable as $\mathbf{v} = \left[\mathbf{q}^{\mathrm{T}}, \mathbf{p}^{\mathrm{T}}\right]^{\mathrm{T}}$, Eq. (3.25), is reduced to the following matrix fashion

$$\dot{\mathbf{v}} = \mathbf{J}\frac{\partial H}{\partial \mathbf{v}} \tag{3.26}$$

where the matrix $\mathbf{J} \in \mathbb{R}^{2n\times 2n}$ is the unit symplectic matrix with the following expression

$$\mathbf{J} = \begin{bmatrix} \mathbf{0}_{n\times n} & \mathbf{I}_{n\times n} \\ -\mathbf{I}_{n\times n} & \mathbf{0}_{n\times n} \end{bmatrix} \tag{3.27}$$

The Hamiltonian system in Eq. (3.25), is a continuous system. However, the continuous system is approximated by a discretization framework when constructing numerical methods. Hence, the property of symplectic conservations is expected for a numerical method to solve Hamiltonian systems such that the symplectic structure of the original continuous system is maintained.

Consider the Hamiltonian system in Eq. (3.25), supposing that state variables at any two adjacent discrete time nodes satisfy the following relationship in a numerical method $\mathcal{A}$

$$\mathbf{v}_{k+1} = \mathbf{\Lambda}(\mathbf{v}_k) \tag{3.28}$$

where $\mathbf{v}_k = \left[\mathbf{q}_k^{\mathrm{T}}, \mathbf{p}_k^{\mathrm{T}}\right]^{\mathrm{T}}$ is the state variable at the $k$ th discrete time node; $\mathbf{\Lambda} : \mathbb{R}^{2n} \to \mathbb{R}^{2n}$ is the mapping function in the state space. Define the Jacobian matrix of the mapping function $\mathbf{\Lambda}$ as

$$\mathbf{K} = \begin{bmatrix} \frac{\partial \mathbf{q}_{k+1}}{\partial \mathbf{q}_k} & \frac{\partial \mathbf{q}_{k+1}}{\partial \mathbf{p}_k} \\ \frac{\partial \mathbf{p}_{k+1}}{\partial \mathbf{q}_k} & \frac{\partial \mathbf{p}_{k+1}}{\partial \mathbf{p}_k} \end{bmatrix} \tag{3.29}$$

If the Jacobian matrix $\mathbf{K}$ satisfies the following condition

$$\mathbf{K}^{\mathrm{T}}\mathbf{J}\mathbf{K} = \mathbf{J} \tag{3.30}$$

Thus, we say that the numerical method $\mathcal{A}$ is symplectic conservative, i.e., the numerical method $\mathcal{A}$ owns the property of symplectic conservation. We can also call $\mathcal{A}$ a symplectic method. For more detail, one can refer to [3].

### 3.4.2 Action and the Least Action Principle

Consider the optimal control problem described in Sect. 3.2.1, for any time interval $[a, b]$ with $t_s \le a < b \le t_f$, the action $S$ within the time interval is defined as

$$S = \int_a^b \left[\boldsymbol{\lambda}^{\mathrm{T}}\dot{\mathbf{x}} - H(\mathbf{x}, \boldsymbol{\lambda}, t)\right]\mathrm{d}t \tag{3.31}$$

According to the least action principle, one has the following variational equation

$$\delta S = \int_a^b \left[\delta\boldsymbol{\lambda}^{\mathrm{T}}\left(\dot{\mathbf{x}} - \frac{\partial H}{\partial \boldsymbol{\lambda}}\right) - \delta\mathbf{x}^{\mathrm{T}}\left(\dot{\lambda} + \frac{\partial H}{\partial \mathbf{x}}\right)\right]\mathrm{d}t + \left(\delta\mathbf{x}^{\mathrm{T}}\boldsymbol{\lambda}\right)\Big|_{t=a}^{t=b} = 0 \tag{3.32}$$

For an optimal control problem, since the Hamiltonian canonical equations (3.23)–(3.24), are satisfied, Eq. (3.32), can be further reduced to

$$\delta S = \delta\mathbf{x}\big|_{t=b}^{\mathrm{T}}\,\boldsymbol{\lambda}\big|_{t=b} - \delta\mathbf{x}\big|_{t=a}^{\mathrm{T}}\,\boldsymbol{\lambda}\big|_{t=a} \tag{3.33}$$

Equation (3.33) suggests that for an optimal control problem, the action within a certain time interval is only the function of state variable at the left boundary (i.e., $\mathbf{x}|_{t=a}$) and the state variable at the right boundary (i.e., $\mathbf{x}|_{t=b}$). To simplify the following derivations, we define the following notations: $\mathbf{x}_a = \mathbf{x}|_{t=a}$, $\mathbf{x}_b = \mathbf{x}|_{t=b}$, $\boldsymbol{\lambda}_a = \boldsymbol{\lambda}|_{t=a}$ and $\boldsymbol{\lambda}_b = \boldsymbol{\lambda}|_{t=b}$.

### *3.4.3 Generating Function*

The most fundamental property of Hamiltonian systems is that the phase flow is a symplectic transformation. To construct symplectic methods for solving optimal control problems, one should guarantee that the state transformation between any two adjacent discrete time nodes is a canonical transformation. The canonical transformation is closely associated with generating functions. In this section, based on the action introduced in Sect. 3.4.2, four kinds of generating functions are considered.

(1) *The first kind of generating function*

If we take the state at the left boundary $\mathbf{x}_a$ and the state at the right boundary $\mathbf{x}_b$ as independent variables, the action $S$ is seen as the first kind of generating function

$$V_1(\mathbf{x}_a, \mathbf{x}_b) = S(\mathbf{x}_a, \mathbf{x}_b) \tag{3.34}$$

Applying the variational principle to $V_1$, one can obtain the relationship similar to Eq. (3.33) as

$$\delta V_1 = \delta\mathbf{x}_b^T\boldsymbol{\lambda}_b - \delta\mathbf{x}_a^T\boldsymbol{\lambda}_a \tag{3.35}$$

which gives the canonical transformation from state variables at two boundaries $\mathbf{x}_a$ and $\mathbf{x}_b$ to costate variables at two boundaries $\boldsymbol{\lambda}_a$ and $\boldsymbol{\lambda}_b$

$$\boldsymbol{\lambda}_a = -\frac{\partial V_1(\mathbf{x}_a, \mathbf{x}_b)}{\partial \mathbf{x}_a}, \ \boldsymbol{\lambda}_b = \frac{\partial V_1(\mathbf{x}_a, \mathbf{x}_b)}{\partial \mathbf{x}_b} \tag{3.36}$$

(2) *The second kind of generating function*

If we take the state at the left boundary $\mathbf{x}_a$ and the costate at the right boundary $\boldsymbol{\lambda}_b$ as independent variables, the second kind of generating function within the time interval is defined as

$$V_2(\mathbf{x}_a, \boldsymbol{\lambda}_b) = \mathbf{x}_b^{\mathrm{T}}\boldsymbol{\lambda}_b - S(\mathbf{x}_a, \mathbf{x}_b) \tag{3.37}$$

Applying the variational principle to $V_2$, one has

$$\delta V_2 = \delta \mathbf{x}_a^T \boldsymbol{\lambda}_a + \mathbf{x}_b^T \delta \boldsymbol{\lambda}_b \tag{3.38}$$

which gives the canonical transformation from the state at the left boundary $\mathbf{x}_a$ and the costate at the right boundary $\boldsymbol{\lambda}_b$ to the costate at the left boundary $\boldsymbol{\lambda}_a$ and the state at the right boundary $\mathbf{x}_b$

$$\boldsymbol{\lambda}_a = \frac{\delta V_2(\mathbf{x}_a, \boldsymbol{\lambda}_b)}{\delta \mathbf{x}_a}, \ \mathbf{x}_b = \frac{\delta V_2(\mathbf{x}_a, \boldsymbol{\lambda}_b)}{\delta \boldsymbol{\lambda}_a} \tag{3.39}$$

(3) *The third kind of generating function*

If we take the costate at the left boundary $\boldsymbol{\lambda}_a$ and the state at the right boundary $\mathbf{x}_b$ as independent variables, the third kind of generating function within the time interval is defined as

$$V_3(\boldsymbol{\lambda}_a, \mathbf{x}_b) = \mathbf{x}_a^{\mathrm{T}}\boldsymbol{\lambda}_a + S(\mathbf{x}_a, \mathbf{x}_b) \tag{3.40}$$

Applying the variational principle to $V_3$, one has

$$\delta V_3 = \mathbf{x}_a^T \delta \boldsymbol{\lambda}_a + \delta \mathbf{x}_b^T \boldsymbol{\lambda}_b \tag{3.41}$$

which gives the canonical transformation from the costate at the left boundary $\boldsymbol{\lambda}_a$ and the state at the right boundary $\mathbf{x}_b$ to the state at the left boundary $\mathbf{x}_a$ and the costate at the right boundary $\boldsymbol{\lambda}_b$

$$\mathbf{x}_a = \frac{\delta V_3(\boldsymbol{\lambda}_a, \mathbf{x}_b)}{\delta \boldsymbol{\lambda}_a}, \ \boldsymbol{\lambda}_b = \frac{\delta V_3(\boldsymbol{\lambda}_a, \mathbf{x}_b)}{\delta \mathbf{x}_b} \tag{3.42}$$

(4) *The fourth kind of generating function*

If we take the costate at the left boundary $\boldsymbol{\lambda}_a$ and the costate at the right boundary $\boldsymbol{\lambda}_b$ as independent variables, the fourth kind of generating function within the time interval is defined as

$$V_4(\boldsymbol{\lambda}_a, \boldsymbol{\lambda}_b) = \mathbf{x}_a^{\mathrm{T}} \boldsymbol{\lambda}_a - \mathbf{x}_b^{\mathrm{T}} \boldsymbol{\lambda}_b + S(\mathbf{x}_a, \mathbf{x}_b) \tag{3.43}$$

Applying the variational principle to $V_4$, one has

$$\delta V_4 = \delta \mathbf{x}_a^{\mathrm{T}} \boldsymbol{\lambda}_a + \mathbf{x}_a^{\mathrm{T}} \delta \boldsymbol{\lambda}_a - \delta \mathbf{x}_b^{\mathrm{T}} \boldsymbol{\lambda}_b - \mathbf{x}_b^{\mathrm{T}} \delta \boldsymbol{\lambda}_b + \delta \mathbf{x}_b^{T} \boldsymbol{\lambda}_b - \delta \mathbf{x}_a^{T} \boldsymbol{\lambda}_a = \mathbf{x}_a^{\mathrm{T}} \delta \boldsymbol{\lambda}_a - \mathbf{x}_b^{\mathrm{T}} \delta \boldsymbol{\lambda}_b \tag{3.44}$$

which gives the canonical transformation from costate variables at two boundaries $\boldsymbol{\lambda}_a$ and $\boldsymbol{\lambda}_b$ to state variables at two boundaries $\mathbf{x}_a$ and $\mathbf{x}_b$

$$\mathbf{x}_a = \frac{\delta V_4(\boldsymbol{\lambda}_a, \boldsymbol{\lambda}_b)}{\delta \boldsymbol{\lambda}_a}, \mathbf{x}_b = -\frac{\delta V_4(\boldsymbol{\lambda}_a, \boldsymbol{\lambda}_b)}{\delta \boldsymbol{\lambda}_b} \tag{3.45}$$

From the constructions of the above four kinds of generating functions, it is seen that different selections of independent variables result in different generating functions. Among the four variables (i.e., $\mathbf{x}_a$, $\mathbf{x}_b$, $\boldsymbol{\lambda}_a$, and $\boldsymbol{\lambda}_b$), any kind of generating function gives the canonical transformation of two variables to another two variables. However, one should be noted that independent variables in four kinds of generating functions always include one variable at the left boundary and another variable at the right boundary.

## 3.5 Legendre Pseudospectral Method

During the construction of symplectic pseudospectral methods, multiple sub-interval Legendre pseudospectral methods are applied to approximate variables in an optimal control problem. This section introduces the Legendre pseudospectral methods. Focusing on the problem in Sect. 3.2.1, we provide how the pseudospectral method is applied to solve a nonlinear optimal control problem. For more detail, interesting readers can refer to [4].

In Sect. 3.5.1, we review the basic theory of Lagrange interpolation and numerical approximation. In Sect. 3.5.2, we introduce the Gauss integration based on the Legendre function. Then in Sect. 3.5.3, definitions of three kinds of Legendre pseudospectral approximation are introduced. The concept of differential matrices in pseudospectral methods is given in Sect. 3.5.4. Finally, in Sect. 3.5.5, we provide how to transform a nonlinear optimal control problem into an NLP by pseudospectral methods.

### 3.5.1 Lagrange Interpolation and Numerical Approximation

For a function $f(x)$, its $N$-degree Lagrange interpolation polynomial $f_N(x)$ is expressed as

$$f_N(x) = \sum_{i=0}^{N} f(x_i)L_i(x) \tag{3.46}$$

where $\{x_k\}_{k=0}^{N}$ are the $(N+1)$ interpolation nodes; $L_k(x)$ with $k = 0, 1, \ldots, N$ represents the Lagrange interpolation basis function corresponding to the $k$th interpolation node, and it is defined as

$$L_k(x) = \prod_{j=0, j\neq k}^{N} \left(\frac{x - x_j}{x_k - x_j}\right), \quad k = 0, 1, \ldots, N \tag{3.47}$$

where the $\{L_k(x)\}_{k=0}^{N}$ satisfy the following orthogonality condition

$$L_k(x_j) = \delta_{kj} = \begin{cases} 1, \ k = j \\ 0, \ k \neq j \end{cases} \tag{3.48}$$

Consider the Legendre polynomial series $\{P_k(x)\}_{k=0}^{\infty}$ defined over $[-1, 1]$. The $k$-degree Legendre polynomial is defined as

$$P_k(x) = \begin{cases} 0, & k = 0 \\ \dfrac{\mathrm{d}^{(k)}}{2^k k! \mathrm{d}x^{(k)}}\left[\left(x^2 - 1\right)^k\right], & \text{otherwise} \end{cases} \tag{3.49}$$

Legendre polynomial series have the orthogonal property, which can be expressed as

$$\int_{-1}^{1} P_k(x)P_j(x)\mathrm{d}x = \frac{2}{2k+1}\delta_{kj} \tag{3.50}$$

For a quadratically integrable function defined over $[-1, 1]$, e.g., $f(x)$, it can be approximated by Legendre polynomial series as follows:

$$f(x) = \sum_{k=0}^{\infty} c_k P_k(x) \tag{3.51}$$

where the coefficients $\{c_k\}_{k=0}^{\infty}$ in Eq. (3.51), can be obtained according to the orthogonality property in Eq. (3.50), as follows:

$$c_k = \frac{2k+1}{2} \int_{-1}^{1} P_k(x) f(x) \mathrm{d}x \tag{3.52}$$

Taking the finite truncation of first $(N+1)$ terms in Eq. (3.51), the $N$-degree approximation of the function $f(x)$ is obtained as follows:

$$f(x) \doteq f^N(x) = \sum_{k=0}^{N} c_k P_k(x) \tag{3.53}$$

### 3.5.2 *Gauss Integration Based on the Legendre Functions*

Consider the numerical integration of function $f(x)$ whose domain is $[-1, 1]$. If the integration points $\{x_k\}_{k=0}^{N}$ are the zeros of a $(N+1)$-degree orthogonal polynomial, then the Gauss integration is expressed as

$$\int_{-1}^{1} f(x) \mathrm{d}x \doteq \sum_{k=0}^{N} w_k f(x_k) \tag{3.54}$$

where $\{w_k\}_{k=0}^{N}$ are the corresponding integration weights. The numerical integration in Eq. (3.54), is precisely satisfied if the degree of $f(x)$ is no more than $(2N+1)$. For Legendre polynomials, there are three types of commonly used Gauss integration points, i.e., the Legendre-Gauss (LG) type, the Legendre-Gauss-Radau (LGR) type, and the Legendre-Gauss-Lobatto (LGL) type. For a Gauss numerical integration with $(N+1)$ integration points, the integration points and corresponding weights are defined as follows:

(1) *LG type Gauss integration*

The integration points $\left\{x_k^{\mathrm{LG}}\right\}_{k=0}^{N}$ are the zeros of the polynomial $P_{N+1}(x)$, and the corresponding weights are

$$w_k^{\mathrm{LG}} = \frac{2}{\left(1 - x_k^2\right) \dot{P}_{N+1}\left(x_k^{\mathrm{LG}}\right)}, \quad k = 0, 1, \ldots, N \tag{3.55}$$

where for a polynomial function with the degree no more than $(2N+1)$, the numerical integration scheme in Eq. (3.54), is precisely satisfied.

(2) *LGR type Gauss integration*

The integration points $\{x_k^{\text{LGR}}\}_{k=0}^{N}$ are the zeros of the polynomial $P_N(x) + P_{N+1}(x)$, and the corresponding weights are

$$\begin{cases} w_0^{\text{LGR}} = \dfrac{2}{(N+1)^2} \\ w_k^{\text{LGR}} = \dfrac{2}{(N+1)^2} \dfrac{1 - x_k^{\text{LGR}}}{\left[P_N\left(x_k^{\text{LGR}}\right)\right]^2}, \quad k = 1, 2, \ldots, N \end{cases} \tag{3.56}$$

where for a polynomial function with the degree no more than $2N$, the numerical integration scheme in Eq. (3.54), is precisely satisfied.

(3) *LGL type Gauss integration*

The integration points $\{x_k^{\text{LGL}}\}_{k=0}^{N}$ are the zeros of the polynomial $\left(1 - x^2\right)\dot{P}_N(x)$, and the corresponding weights are

$$w_k^{\text{LGL}} = \frac{2}{N(N+1)} \frac{1}{\left[P_N\left(x_k^{\text{LGL}}\right)\right]^2}, \quad k = 0, 1, \ldots, N \tag{3.57}$$

where for a polynomial function with the degree no more than $(2N-1)$, the numerical integration scheme in Eq. (3.54), is precisely satisfied.

### 3.5.3 Legendre Pseudospectral Approximation

Consider the calculation of coefficients in the function approximation as given in Eq. (3.52), one can use the Gauss integration as follows:

$$c_k = \frac{2k+1}{2} \int_{-1}^{1} P_k(x) f(x) \mathrm{d}x = \frac{2k+1}{2} \sum_{j=0}^{N} P_k\left(x_j^{(\cdot)}\right) f\left(x_j^{(\cdot)}\right) w_j^{(\cdot)} \tag{3.58}$$

where the superscript $(\cdot)$ represents a certain one of LG, LGR or LGL type Gauss integration.

If we take the integration points in any of the three types of Gauss integration as the interpolation nodes to construct the Lagrange interpolation, we get the corresponding Legendre pseudospectral approximation. And the Lagrange interpolation basis functions for these three types of approximation are as follows:

(1) *LG type pseudospectral approximation*

$$L_k^{\mathrm{LG}}(x) = \frac{P_{N+1}(x)}{\left(x - x_k^{\mathrm{LG}}\right)\dot{P}_{N+1}(x)}, \quad k = 0, 1, \ldots, N \tag{3.59}$$

(2) *LGR type pseudospectral approximation*

$$L_k^{\mathrm{LGR}}(x) = \frac{1 - x_k^{\mathrm{LGR}}}{2(N+1)P_N\left(x_k^{\mathrm{LGR}}\right)} \frac{P_N(x) + P_{N+1}(x)}{x - x_k^{\mathrm{LGR}}}, \quad k = 0, 1, \ldots, N \tag{3.60}$$

(3) *LGL type pseudospectral approximation*

$$L_k^{\mathrm{LGL}}(x) = \frac{1}{N(N+1)P_N\left(x_k^{\mathrm{LGL}}\right)} \frac{\left(x^2 - 1\right)\dot{P}_N(x)}{x - x_k^{\mathrm{LGL}}}, \quad k = 0, 1, \ldots, N \tag{3.61}$$

It is noted zero, one, and two boundary points (i.e., the boundary of time domain $[-1, 1]$) are included in the interpolation node list in the LG, LGR, and LGL type pseudospectral approximation, respectively.

***Remark 3.2*** The initial and terminal boundary conditions are usually prescribed in optimal control problems. If the boundary point is included in the interpolation node list, it would be convenient to impose the boundary conditions. In the LGL type pseudospectral approximation, there are interpolation nodes distributed at both boundary points. Hence, the LGL type pseudospectral approximation is adopted for the construction of the symplectic pseudospectral methods in this book. As for the LG and the LGR type pseudospectral approximation, the value of variables at the boundary can be obtained by interpolation. □

### 3.5.4 Differentiation Matrix

For a function $f(x) : \mathbb{R} \to \mathbb{R}$ defined over $[-1, 1]$, its $N$-degree Legendre pseudospectral approximation is

$$f(x) \doteq f^N(x) = \sum_{k=0}^{N} L_k^{(\cdot)}(x) f\left(x_k^{(\cdot)}\right) \tag{3.62}$$

The derivative of $f(x)$ at the interpolation node $x_k^{(\cdot)}$ thus can be approximated by the derivative of $f^N(x)$ at that node

$$\dot{f}\left(x_k^{(\cdot)}\right) \doteq \dot{f}^N\left(x_k^{(\cdot)}\right) = \sum_{j=0}^{N} \dot{L}_j^{(\cdot)}\left(x_k^{(\cdot)}\right) f\left(x_j^{(\cdot)}\right) = \sum\nolimits_{j=0}^{N} D_{kj}^{(\cdot)} f\left(x_j^{(\cdot)}\right) \tag{3.63}$$

where $D_{kj}$ is the value of the derivative of $L_j^{(\cdot)}(x)$ at $x_k^{(\cdot)}$

$$D_{kj} = \left.\frac{\mathrm{d}L_j^{(\cdot)}(x)}{\mathrm{d}x}\right|_{x_k^{(\cdot)}} \tag{3.64}$$

According to Eq. (3.64), the derivatives of $f(x)$ at all interpolation nodes is associated with $\left\{f\left(x_k^{(\cdot)}\right)\right\}_{k=0}^{N}$ via the matrix $\mathbf{D}^{(\cdot)} = \left[D_{kj}^{(\cdot)}\right] \in \mathbb{R}^{(N+1)\times(N+1)}$ as follows:

$$\frac{1}{\mathrm{d}x}\begin{bmatrix} \mathrm{d}f\left(x_0^{(\cdot)}\right) \\ \mathrm{d}f\left(x_1^{(\cdot)}\right) \\ \vdots \\ \mathrm{d}f\left(x_N^{(\cdot)}\right) \end{bmatrix} = \begin{bmatrix} D_{00}^{(\cdot)} & D_{01}^{(\cdot)} & \cdots & D_{0N}^{(\cdot)} \\ D_{10}^{(\cdot)} & D_{11}^{(\cdot)} & \cdots & D_{1N}^{(\cdot)} \\ \vdots & \vdots & \ddots & \vdots \\ D_{N0}^{(\cdot)} & D_{N1}^{(\cdot)} & \cdots & D_{NN}^{(\cdot)} \end{bmatrix} \begin{bmatrix} f\left(x_0^{(\cdot)}\right) \\ f\left(x_1^{(\cdot)}\right) \\ \vdots \\ f\left(x_N^{(\cdot)}\right) \end{bmatrix} = \mathbf{D}^{(\cdot)} \begin{bmatrix} f\left(x_0^{(\cdot)}\right) \\ f\left(x_1^{(\cdot)}\right) \\ \vdots \\ f\left(x_N^{(\cdot)}\right) \end{bmatrix} \tag{3.65}$$

where $\mathbf{D}^{(\cdot)}$ is also called the first-order differentiation matrix and is expressed as follows:

(1) *LG type pseudospectral approximation*

$$D_{kj}^{\mathrm{LG}} = \begin{cases} \frac{\dot{P}_{N+1}\left(x_k^{\mathrm{LG}}\right)}{\dot{P}_{N+1}\left(x_j^{\mathrm{LG}}\right)} \frac{1}{\left(x_k^{\mathrm{LG}} - x_j^{\mathrm{LG}}\right)} & k \neq j \\ \frac{x_k^{\mathrm{LG}}}{1 - \left(x_k^{\mathrm{LG}}\right)^2} & k = j \end{cases} \tag{3.66}$$

(2) *LGR type pseudospectral approximation*

$$D_{kj}^{\text{LGR}} = \begin{cases} -\frac{N(N+2)}{4} & , k = j = 0 \\ \frac{x_k^{\text{LGR}}}{1-(x_k^{\text{LGR}})^2} + \frac{(N+1)P_N(x_k^{\text{LGR}})}{\left(1-(x_k^{\text{LGR}})^2\right)(\dot{P}_{N+1}(x_k^{\text{LGR}})+\dot{P}_N(x_k^{\text{LGR}}))} & , 1 \le k = j \le N \\ \frac{\dot{P}_{N+1}(x_k^{\text{LGR}})+\dot{P}_N(x_k^{\text{LGR}})}{\dot{P}_{N+1}(x_j^{\text{LGR}})+\dot{P}_N(x_j^{\text{LGR}})} \frac{1}{(x_k^{\text{LGR}}-x_j^{\text{LGR}})} & , k \ne j \end{cases} \tag{3.67}$$

(3) *LGL type pseudospectral approximation*

$$D_{kj}^{\text{LGL}} = \begin{cases} -\frac{N(N+1)}{4} & , k = j = 0 \\ \frac{P_N(x_k^{\text{LGL}})}{P_N(x_j^{\text{LGL}})} \frac{1}{(x_k^{\text{LGL}}-x_j^{\text{LGL}})} & , k \ne j, 0 \le k, j \le N \\ 0 & , 1 \le k = j \le N-1 \\ \frac{N(N+1)}{4} & , k = j = N \end{cases} \tag{3.68}$$

Besides, for the $m$th order differentiation matrix $\mathbf{D}^{[m](\cdot)}$, we have $\mathbf{D}^{[m](\cdot)} = \left(\mathbf{D}^{(\cdot)}\right)^m$, that is to say

$$\frac{1}{\mathrm{d}x^m}\begin{bmatrix} \mathrm{d}^m f\left(x_0^{(\cdot)}\right) \\ \mathrm{d}^m f\left(x_1^{(\cdot)}\right) \\ \vdots \\ \mathrm{d}^m f\left(x_N^{(\cdot)}\right) \end{bmatrix} = \mathbf{D}^{[m](\cdot)} \begin{bmatrix} f\left(x_0^{(\cdot)}\right) \\ f\left(x_1^{(\cdot)}\right) \\ \vdots \\ f\left(x_N^{(\cdot)}\right) \end{bmatrix} = \left(\mathbf{D}^{(\cdot)}\right)^m \begin{bmatrix} f\left(x_0^{(\cdot)}\right) \\ f\left(x_1^{(\cdot)}\right) \\ \vdots \\ f\left(x_N^{(\cdot)}\right) \end{bmatrix} \tag{3.69}$$

Among the high order differentiation matrices, the second-order differentiation matrix plays an important role. For example, the second-order differentiation matrix is required to develop adaptive mesh refinement techniques in [5].

### 3.5.5 Legendre Pseudospectral Method for Solving Optimal Control Problems

Without loss of generality, we use the Legendre pseudospectral method to solve the nonlinear optimal control problem in Eqs. (3.1)–(3.3).

To apply the pseudospectral method, the time domain $\left[t_s, t_f\right]$ is usually transformed into a standard interval $[-1, 1]$ by a linear transformation

$$\tau = \frac{2}{t_f - t_s}\left(t - \frac{t_f + t_s}{2}\right) \tag{3.70}$$

Correspondingly, we can obtain the following relationship between integration and differentiation

$$\int_{t_s}^{t_f} (\cdot)\mathrm{d}t = \frac{t_f - t_s}{2} \int_{-1}^{1} (\cdot)\mathrm{d}\tau \tag{3.71}$$

$$\frac{\mathrm{d}}{\mathrm{d}t}(\cdot) = \frac{2}{t_f - t_s} \frac{\mathrm{d}}{\mathrm{d}\tau}(\cdot) \tag{3.72}$$

Apply the Lagrange interpolation to state variables $\mathbf{x}(\tau)$, thus state variables are approximated as follows:

$$\mathbf{x}(\tau) \doteq \sum_{k=0}^{N_{\mathbf{X}}} \mathbf{x}\left(\tau_k^{\mathbf{X}(\cdot)}\right) L_k^{\mathbf{X}(\cdot)}(\tau) \tag{3.73}$$

where $N_{\mathbf{X}}$ is the approximation degree; $\left\{\tau_k^{\mathbf{X}(\cdot)}\right\}_{k=0}^{N_X}$ represents the interpolation node list; $\left\{L_k^{\mathbf{X}(\cdot)}\right\}_{k=0}^{N_X}$ are the Lagrange interpolation basis functions.

Similarly, we can apply the Lagrange interpolation to control variables

$$\mathbf{u}(\tau) \doteq \sum\nolimits_{k=0}^{N_{\mathbf{U}}} \mathbf{u}\left(\tau_k^{\mathbf{U}(\cdot)}\right) L_k^{\mathbf{U}(\cdot)}(\tau) \tag{3.74}$$

where $N_{\mathbf{U}}$, $\left\{\tau_k^{\mathbf{U}(\cdot)}\right\}_{k=0}^{N_U}$ and $\left\{L_k^{\mathbf{U}(\cdot)}\right\}_{k=0}^{N_U}$ have similar meanings as those in Eq. (3.73).

***Remark 3.3*** The Lagrange interpolation nodes for state variables $\tau_k^{\mathbf{X}(\cdot)}$ are not necessarily identical to the corresponding type of Gauss integration points. As stated in Remark 3.2, to facilitate the imposing of boundary conditions in LG and LGR type pseudospectral methods, the interval boundary is usually also included in the node list; and the Lagrange interpolation basis function and first-order differentiation matrix should be modified correspondingly. □

It is seen that the first-order derivative of state variables is required in Eq. (3.2). Hence, according to Sect. 3.5.4, one has

$$\dot{\mathbf{x}}\left(\tau_k^{(\cdot)}\right) \doteq \sum_{j=0}^{N} \dot{L}_j^{\mathbf{X}(\cdot)}(\tau)\mathbf{x}\left(\tau_j^{\mathbf{X}(\cdot)}\right) = \sum_{j=0}^{N} D_{kj}^{\mathbf{X}(\cdot)}\mathbf{x}\left(\tau_j^{\mathbf{X}(\cdot)}\right) \tag{3.75}$$

Based on the above approximation, pseudospectral methods transform the optimal control problem in Eqs. (3.1)–(3.3), into the following nonlinear programming problem: Find the optimal control sequence $\left\{\mathbf{u}\left(\tau_k^{(\cdot)}\right)\right\}_{k=0}^{N_U}$ and corresponding state trajectory $\left\{\mathbf{x}\left(\tau_k^{\mathbf{X}(\cdot)}\right)\right\}_{k=0}^{N_{\mathbf{X}}}$, such that minimizing the following discretized cost functional

$$J_N = \Psi(\mathbf{x}(1)) + \frac{t_f - t_s}{2}\int_{-1}^{1} \Phi(\mathbf{x}(\tau), \mathbf{u}(\tau), \tau)\mathrm{d}\tau$$
$$= \Psi(\mathbf{x}(1)) + \frac{t_f - t_s}{2}\sum_{k=0}^{N} w_k^{(\cdot)} \Phi\left(\mathbf{x}\left(\tau_k^{(\cdot)}\right), \mathbf{u}\left(\tau_k^{(\cdot)}\right), t_k^{(\cdot)}\right) \quad (3.76)$$

subject to discretized system equations

$$\sum_{j=0}^{N_{\mathbf{X}}} D_{kj}^{\mathbf{X}(\cdot)} \mathbf{x}\left(\tau_j^{\mathbf{X}(\cdot)}\right) = \frac{t_f - t_s}{2}\mathbf{f}\left(\mathbf{x}\left(\tau_k^{(\cdot)}\right), \mathbf{u}\left(\tau_k^{(\cdot)}\right), t_k^{(\cdot)}\right), \quad k = 0, 1, \ldots, N \quad (3.77)$$

and initial conditions

$$\mathbf{x}(-1) = \mathbf{x}_s \quad (3.78)$$

Equations (3.76)–(3.78) constitute a finite-dimensional nonlinear programming problem. Equation (3.76) is the objective function, Eqs. (3.77)–(3.78), are the constraints, and $\left\{\mathbf{u}\left(\tau_k^{(\cdot)}\right)\right\}_{k=0}^{N}$ and $\left\{\mathbf{x}\left(\tau_k^{(\cdot)}\right)\right\}_{k=0}^{N}$ are the variables to be optimized. Such a nonlinear programming problem can be solved efficiently by NLP solvers such as IPOPT [6] and SNOPT [7].

## References


1. Bryson AE, Ho YC (1975) Applied optimal control. Hemisphere/Wiley, New York
2. Gergaud J, Haberkorn T (2006) Homotopy method for minimum consumption orbit transfer problem. ESAIM: Control Optim Calc Var 12(2):294–310
3. Zhong W (2004) Duality system in applied mechanics and optimal control. Springer, Boston
4. Shen J, Tang T, Wang L (2011) Spectral methods: algorithms, analysis and applications. Springer, Berlin
5. Peng H, Wang X, Li M et al (2017) An hp symplectic pseudospectral method for nonlinear optimal control. Commun Nonlinear Sci Numer Simul 42:623–644
6. Wächter A, Biegler LT (2006) On the implementation of an interior-point filter line-search algorithm for large-scale nonlinear programming. Math Program 106(1):25–57
7. Gill PE, Murray W, Saunders MA (2005) SNOPT: an SQP algorithm for large-scale constrained optimization. SIAM Rev 47(1):99–131

# Chapter 4
# SPM for Unconstrained Nonlinear Optimal Control Problems

## 4.1 Introduction

Problem without constraints is the simplest formulation of nonlinear optimal control problems. When novel numerical methods are proposed, they are frequently tested by unconstrained problems first. As the first chapter in Part II of the book, we also start with unconstrained problems. A symplectic pseudospectral method based on the first kind of generating function is developed. Furthermore, a mesh refinement technique based on the relative curvature of state variables is presented to achieve faster convergence. Readers interested in the proof of the symplectic conservation property of the algorithm can refer to [1].

## 4.2 Problem Formulation

Consider the following unconstrained nonlinear optimal control problem:

(Problem 4)

Minimize

$$J = \Psi\left(\mathbf{x}\left(t_f\right)\right) + \int_{t_s}^{t_f} \Phi(\mathbf{x}(t), \mathbf{u}(t), t)\mathrm{d}t \tag{4.1}$$

subject to the system equation and initial conditions



X. Wang et al., *Symplectic Pseudospectral Methods for Optimal Control*, Intelligent Systems, Control and Automation: Science and Engineering 97, https://doi.org/10.1007/978-981-15-3438-6_4

$$\dot{\mathbf{x}} = \mathbf{f}(\mathbf{x}, \mathbf{u}, t),\ \mathbf{x}(t_s) = \mathbf{x}_s \tag{4.2}$$

where $t_s$ and $t_f$ are the fixed initial and terminal time, respectively.

Two types of terminal boundary conditions, i.e., the fixed terminal state and the free terminal states are considered. And as for the condition of the fixed terminal state, the prescribed terminal state is denoted as $\mathbf{x}_g \in \mathbb{R}^d$.

Then, according to Sect. 3.3, the first-order necessary conditions of Problem 4 can be derived. First, by introducing the costate variable $\boldsymbol{\lambda}$, the Hamiltonian of Problem 4 can be expressed as

$$\bar{H}(\mathbf{x}, \boldsymbol{\lambda}, \mathbf{u}, t) = \Phi(\mathbf{x}, \mathbf{u}, t) + \boldsymbol{\lambda}^{\mathrm{T}}\mathbf{f}(\mathbf{x}, \mathbf{u}, t) \tag{4.3}$$

Then according to the optimality condition with respect to the control variable, one has

$$\frac{\partial \bar{H}(\mathbf{x}, \boldsymbol{\lambda}, \mathbf{u}, t)}{\partial \mathbf{u}} = \mathbf{0} \tag{4.4}$$

By solving Eq. (4.4), we can express the control variable as a function of state and costate variables as follow

$$\mathbf{u} = \mathbf{u}(\mathbf{x}, \boldsymbol{\lambda}, t) \tag{4.5}$$

Substituting Eq. (4.5), back into Eq. (4.3), the Hamiltonian can be written as the function of state and costate variables, i.e., $H(\mathbf{x}, \boldsymbol{\lambda}, t)$. The optimality conditions with reapect to state and costate variables, which are also called the Hamiltonian canonical equations, can be derived as

$$\dot{\mathbf{x}} = \frac{\partial H(\mathbf{x}, \boldsymbol{\lambda}, t)}{\partial \boldsymbol{\lambda}} = \mathbf{f}(\mathbf{x}, \mathbf{u}, t) \tag{4.6}$$

$$\dot{\boldsymbol{\lambda}} = -\frac{\partial H(\mathbf{x}, \boldsymbol{\lambda}, t)}{\partial \mathbf{x}} \tag{4.7}$$

It's noted that the terminal state can be either fixed or free, thus the terminal boundary condition can be derived according to Sect. 3.3. For a fixed terminal state, one has

$$\mathbf{x}(t_s) = \mathbf{x}_s,\ \mathbf{x}(t_f) = \mathbf{x}_g \tag{4.8}$$

and for the free terminal state, one has

$$\mathbf{x}(t_s) = \mathbf{x}_s,\ \boldsymbol{\lambda}(t_f) = \frac{\partial \Psi(\mathbf{x}(t_f))}{\partial \mathbf{x}(t_f)} \tag{4.9}$$

Now, the first-order necessary conditions and the boundary conditions for Problem 4 are already derived. It's seen that the first-order necessary conditions for this problem are a Hamiltonian two-point boundary value problem.

## 4.3 Algorithm Construction

Consider the first-order necessary conditions derived in Sect. 4.2, based on the first kind of generating function and the LGL local pseudospectral method, a symplectic pseudospectral method to solve Problem 4 is developed. The SPM is mainly divided into 5 steps:

(1) Dividing the time domain into several sub-intervals.
(2) Applying the LGL pseudospectral method to state and costate variables in each sub-interval.
(3) Applying the variational principle to the first kind of generating function to form a system of nonlinear algebraic equations.
(4) Applying the boundary conditions to the system of nonlinear algebraic equations.
(5) Solving the system of nonlinear algebraic equations by the Newton–Raphson method.

### 4.3.1 *Time Domain Discretization*

Divide the time domain $\left[t_s, t_f\right]$ into $Q$ sub-intervals, and the $j$ th sub-interval is denoted as $\Gamma^{(j)} = \left[t_{j-1}, t_j\right]$, $j = 1, 2, \ldots, Q$, where $t_s = t_0 < t_1 < \ldots < t_Q = t_f$.

### 4.3.2 *Applying the LGL Pseudospectral Method*

To apply the LGL pseudospectral method in each sub-interval, the $j$th sub-interval is transformed into a standard time interval $[-1, 1]$ by the following linear transformation

$$\tau^{(j)} = \frac{2}{t_j - t_{j-1}}\left(t - \frac{t_j + t_{j-1}}{2}\right) \tag{4.10}$$

Thus, the first kind of generating function of the $j$th sub-interval can be written as

$$S^{(j)} = \int_{-1}^{1} \left( \boldsymbol{\lambda}^{\mathrm{T}} \frac{\mathrm{d}\mathbf{x}}{\mathrm{d}\tau} - \frac{t_j - t_{j-1}}{2} H(\mathbf{x}, \boldsymbol{\lambda}, \tau) \right) \mathrm{d}\tau \tag{4.11}$$

In the $j$th sub-interval, state and costate variables are approximated by the $N^{(j)}$-order LGL pseudospectral method as follows:

$$\mathbf{x}^{(j)}(\tau) = \sum_{l=0}^{N^{(j)}} \mathbf{x}_l^{(j)} \rho_l^{(j)}(\tau), \ \boldsymbol{\lambda}^{(j)}(\tau) = \sum_{l=0}^{N^{(j)}} \boldsymbol{\lambda}_l^{(j)} \rho_l^{(j)}(\tau) \tag{4.12}$$

As in symplectic methods based on the first kind of generating function, for any sub-interval, states at the left and the right ends are taken as independent variables. Hence, for the simplicity of the following derivation, we denote the non-independent variables within the $j$th sub-interval by following notations

$$\bar{\mathbf{x}}^{(j)} = \left[ \left(\mathbf{x}_1^{(j)}\right)^{\mathrm{T}}, \left(\mathbf{x}_2^{(j)}\right)^{\mathrm{T}}, \ldots, \left(\mathbf{x}_{N^{(j)}-1}^{(j)}\right)^{\mathrm{T}} \right]^{\mathrm{T}} \tag{4.13}$$

$$\bar{\boldsymbol{\lambda}}^{(j)} = \left[ \left(\boldsymbol{\lambda}_0^{(j)}\right)^{\mathrm{T}}, \left(\boldsymbol{\lambda}_1^{(j)}\right)^{\mathrm{T}}, \ldots, \left(\boldsymbol{\lambda}_{N^{(j)}}^{(j)}\right)^{\mathrm{T}} \right]^{\mathrm{T}} \tag{4.14}$$

Hence, the first kind of generating function of the $j$th sub-interval can be expressed in the following fashion by the LGL type Gauss integration

$$S^{(j)} = \sum_{k=0}^{N^{(j)}} w_k^{(j)} \left( \left(\boldsymbol{\lambda}_k^{(j)}\right)^{\mathrm{T}} \sum_{l=0}^{N^{(j)}} D_{kl}^{(j)} \mathbf{x}_l^{(j)} - \frac{t_j - t_{j-1}}{2} H_k^{(j)} \right) \tag{4.15}$$

where $H_k^{(j)}$ represents the value of Hamiltonian at $\tau_k^{(j)}$; $w_k^{(j)}$ is the integral weight as defined in Eq. (3.57); and $\left[D_{kl}^{(j)}\right]$ is the first-order differentiation matrix defined in Eq. (3.68).

### 4.3.3 *Applying the Variational Principle to the First Kind of Generating Function*

According to Eq. (3.35), one has the following stationary conditions

$$\frac{\partial S^{(j)}\left(\mathbf{x}_{j-1}, \bar{\mathbf{x}}^{(j)}, \mathbf{x}_j, \bar{\boldsymbol{\lambda}}^{(j)}\right)}{\partial \bar{\mathbf{x}}^{(j)}} = \mathbf{0} \tag{4.16}$$

$$\frac{\partial S^{(j)}\left(\mathbf{x}_{j-1}, \bar{\mathbf{x}}^{(j)}, \mathbf{x}_j, \bar{\lambda}^{(j)}\right)}{\partial \bar{\lambda}^{(j)}} = \mathbf{0} \tag{4.17}$$

By summing up the action of each sub-interval, the action of the time interval $[t_s, t_f]$ is obtained as follows:

$$S(\mathbf{x}_0, \mathbf{x}_Q) = \sum_{j=1}^{Q} S^{(j)}(\mathbf{x}_{j-1}, \mathbf{x}_j) \tag{4.18}$$

Applying Eq. (3.35) to Eq. (4.18), one has the following stationary conditions

$$\frac{\partial S^{(j)}}{\partial \mathbf{x}_j} + \frac{\partial S^{(j+1)}}{\partial \mathbf{x}_j} = 0, \quad j = 1, 2, \ldots, Q-1 \tag{4.19}$$

Additionally, according to Eq. (3.36), the costate at $t = t_s$ and $t = t_f$ can be obtained as follows:

$$\boldsymbol{\lambda}_0 = -\frac{\partial S(\mathbf{x}_0, \mathbf{x}_Q)}{\partial \mathbf{x}_0} = -\frac{\partial S^{(1)}(\mathbf{x}_0, \mathbf{x}_1)}{\partial \mathbf{x}_0} \tag{4.20}$$

$$\boldsymbol{\lambda}_Q = \frac{\partial S(\mathbf{x}_0, \mathbf{x}_Q)}{\partial \mathbf{x}_Q} = \frac{\partial S^{(Q)}(\mathbf{x}_{Q-1}, \mathbf{x}_Q)}{\partial \mathbf{x}_Q} \tag{4.21}$$

It is seen that Problem 4 is transformed into a system of nonlinear algebraic equations, i.e., Equations (4.16)–(4.17) and Eqs. (4.19)–(4.21). To solve the system of nonlinear algebraic equations by the Newton–Raphson method efficiently, the explicit expression of the Jacobian matrix is derived.

For the $j$th sub-interval, the partial derivative of the first kind of generating function with respect to state and costate variables can be expressed as follows:

$$\left(\mathbf{f}_0^x\right)^{(j)} = -\boldsymbol{\lambda}_{j-1} \tag{4.22}$$

$$\left(\mathbf{f}_m^x\right)^{(j)} = \mathbf{0}, \quad m = 1, 2, 3, \ldots, N^{(j)} - 1 \tag{4.23}$$

$$\left(\mathbf{f}_{N^{(j)}}^x\right)^{(j)} = \boldsymbol{\lambda}_j \tag{4.24}$$

$$\left(\mathbf{f}_m^\lambda\right)^{(j)} = \mathbf{0}, \quad m = 0, 1, 2, \ldots, N^{(j)} \tag{4.25}$$

where

$$\left(\mathbf{f}_m^x\right)^{(j)} = \frac{\partial S^{(j)}}{\partial \mathbf{x}_m^{(j)}} = \sum_{k=0}^{N^{(j)}} w_k^{(j)} D_{km}^{(j)} \boldsymbol{\lambda}_k^{(j)} - w_m^{(j)} \frac{t_j - t_{j-1}}{2} \cdot \frac{\partial H\left(\mathbf{x}_m^{(j)}, \boldsymbol{\lambda}_m^{(j)}, \theta_m^{(j)}\right)}{\partial \mathbf{x}_m^{(j)}} \tag{4.26}$$

$$\left(\mathbf{f}_m^\lambda\right)^{(j)} = \frac{\partial S^{(j)}}{\partial \boldsymbol{\lambda}_m^{(j)}} = w_m^{(j)} \left( \sum_{k=0}^{N^{(j)}} D_{mk}^{(j)} \mathbf{x}_k^{(j)} - \frac{t_j - t_{j-1}}{2} \cdot \frac{\partial H\left(\mathbf{x}_m^{(j)}, \boldsymbol{\lambda}_m^{(j)}, \theta_m^{(j)}\right)}{\partial \boldsymbol{\lambda}_m^{(j)}} \right) \tag{4.27}$$

By taking the partial derivatives of Eqs. (4.26)–(4.27), with respect to state and costate variables, the information on the Jacobian matrix related to the $j$th sub-interval can be obtained

$$\left(\mathbf{K}_{mn}^{xx}\right)^{(j)} = \frac{\left(\partial \mathbf{f}_m^x\right)^{(j)}}{\partial \mathbf{x}_n^{(j)}} = -\frac{t_j - t_{j-1}}{2} \cdot \frac{\partial^2 H\left(\mathbf{x}_m^{(j)}, \boldsymbol{\lambda}_m^{(j)}, \theta_m^{(j)}\right)}{\partial \mathbf{x}_m^{(j)} \partial \mathbf{x}_n^{(j)}} \delta_m^n \tag{4.28}$$

$$\left(\mathbf{K}_{mn}^{x\lambda}\right)^{(j)} = \frac{\left(\partial \mathbf{f}_m^x\right)^{(j)}}{\partial \boldsymbol{\lambda}_n^{(j)}} = w_n^{(j)} D_{mn}^{(j)} \mathbf{I} - w_m^{(j)} \frac{t_j - t_{j-1}}{2} \cdot \frac{\partial^2 H\left(\mathbf{x}_m^{(j)}, \boldsymbol{\lambda}_m^{(j)}, \theta_m^{(j)}\right)}{\partial \mathbf{x}_m^{(j)} \partial \boldsymbol{\lambda}_n^{(j)}} \delta_m^n \tag{4.29}$$

$$\begin{aligned}\left(\mathbf{K}_{mn}^{\lambda x}\right)^{(j)} &= \frac{\left(\partial \mathbf{f}_m^\lambda\right)^{(j)}}{\partial \mathbf{x}_n^{(j)}} = w_m^{(j)} D_{mn}^{(j)} \mathbf{I} - w_m^{(j)} \frac{t_j - t_{j-1}}{2} \cdot \frac{\partial^2 H\left(\mathbf{x}_m^{(j)}, \boldsymbol{\lambda}_m^{(j)}, \theta_m^{(j)}\right)}{\partial \boldsymbol{\lambda}_m^{(j)} \partial \mathbf{x}_n^{(j)}} \delta_m^n \\ &= \left(\left(K_{nm}^{x\lambda}\right)^{(j)}\right)^{\mathrm{T}}\end{aligned} \tag{4.30}$$

$$\left(\mathbf{K}_{mn}^{\lambda\lambda}\right)^{(j)} = \frac{\left(\partial \mathbf{f}_m^\lambda\right)^{(j)}}{\partial \boldsymbol{\lambda}_n^{(j)}} = -w_m^{(j)} \frac{t_j - t_{j-1}}{2} \cdot \frac{\partial^2 H\left(\mathbf{x}_m^{(j)}, \boldsymbol{\lambda}_m^{(j)}, \theta_m^{(j)}\right)}{\partial \boldsymbol{\lambda}_m^{(j)} \partial \boldsymbol{\lambda}_n^{(j)}} \delta_m^n \tag{4.31}$$

For the system of nonlinear algebraic equations, we denote the right-hand side related to the $j$th sub-interval as the $\mathbf{f}^{(j)}$, and denote the Jacobian matrix related to the $j$th sub-interval as $\mathbf{K}^{(j)}$. The expressions of such two variables are as follows:

$$
\begin{aligned}
\mathbf{K}^{(j)} &= \begin{bmatrix} \mathbf{K}_{11}^{(j)} & \mathbf{K}_{12}^{(j)} & \mathbf{K}_{13}^{(j)} & \mathbf{K}_{14}^{(j)} \\ \mathbf{K}_{21}^{(j)} & \mathbf{K}_{22}^{(j)} & \mathbf{K}_{23}^{(j)} & \mathbf{K}_{24}^{(j)} \\ \mathbf{K}_{31}^{(j)} & \mathbf{K}_{32}^{(j)} & \mathbf{K}_{33}^{(j)} & \mathbf{K}_{34}^{(j)} \\ \mathbf{K}_{41}^{(j)} & \mathbf{K}_{42}^{(j)} & \mathbf{K}_{43}^{(j)} & \mathbf{K}_{44}^{(j)} \end{bmatrix} \\
&= \left[\begin{array}{c|ccc|c|cccc}
\left(\mathbf{K}_{00}^{xx}\right)^{(j)} & \mathbf{0}_d & \cdots & \mathbf{0}_d & \mathbf{0}_d & \left(\mathbf{K}_{00}^{x\lambda}\right)^{(j)} & \left(\mathbf{K}_{01}^{x\lambda}\right)^{(j)} & \cdots \left(\mathbf{K}_{0,N^{(j)}-1}^{x\lambda}\right)^{(j)} & \left(\mathbf{K}_{0N^{(j)}}^{x\lambda}\right)^{(j)} \\ \hline
\mathbf{0}_d & \left(\mathbf{K}_{11}^{xx}\right)^{(j)} & \cdots & \mathbf{0}_d & \mathbf{0}_d & \left(\mathbf{K}_{10}^{x\lambda}\right)^{(j)} & \left(\mathbf{K}_{11}^{x\lambda}\right)^{(j)} & \cdots \left(\mathbf{K}_{1,N^{(j)}-1}^{x\lambda}\right)^{(j)} & \left(\mathbf{K}_{1N^{(j)}}^{x\lambda}\right)^{(j)} \\
\vdots & \vdots & \ddots & \vdots & \vdots & \vdots & \vdots & \ddots \quad \vdots & \vdots \\
\mathbf{0}_d & \mathbf{0}_d & \cdots & \left(\mathbf{K}_{N^{(j)}-1,N^{(j)}-1}^{xx}\right)^{(j)} & \mathbf{0}_d & \left(\mathbf{K}_{N^{(j)}-1,0}^{x\lambda}\right)^{(j)} & \left(\mathbf{K}_{N^{(j)}-1,1}^{x\lambda}\right)^{(j)} & \cdots \left(\mathbf{K}_{N^{(j)}-1,N^{(j)}-1}^{x\lambda}\right)^{(j)} & \left(\mathbf{K}_{N^{(j)}-1,N^{(j)}}^{x\lambda}\right)^{(j)} \\ \hline
\mathbf{0}_d & \mathbf{0}_d & \cdots & \mathbf{0}_d & \left(\mathbf{K}_{N^{(j)}N^{(j)}}^{xx}\right)^{(j)} & \left(\mathbf{K}_{N^{(j)}0}^{x\lambda}\right)^{(j)} & \left(\mathbf{K}_{N^{(j)}1}^{x\lambda}\right)^{(j)} & \cdots \left(\mathbf{K}_{N^{(j)},N^{(j)}-1}^{x\lambda}\right)^{(j)} & \left(\mathbf{K}_{N^{(j)}N^{(j)}}^{x\lambda}\right)^{(j)} \\ \hline
\left(\mathbf{K}_{00}^{\lambda x}\right)^{(j)} & \left(\mathbf{K}_{01}^{\lambda x}\right)^{(j)} & \cdots & \left(\mathbf{K}_{0,N^{(j)}-1}^{\lambda x}\right)^{(j)} & \left(\mathbf{K}_{0N^{(j)}}^{\lambda x}\right)^{(j)} & \left(\mathbf{K}_{00}^{\lambda\lambda}\right)^{(j)} & \mathbf{0}_d & \cdots \quad \mathbf{0}_d & \mathbf{0}_d \\
\left(\mathbf{K}_{10}^{\lambda x}\right)^{(j)} & \left(\mathbf{K}_{11}^{\lambda x}\right)^{(j)} & \cdots & \left(\mathbf{K}_{1,N^{(j)}-1}^{\lambda x}\right)^{(j)} & \left(\mathbf{K}_{1N^{(j)}}^{\lambda x}\right)^{(j)} & \mathbf{0}_d & \left(\mathbf{K}_{11}^{\lambda\lambda}\right)^{(j)} & \cdots \quad \mathbf{0}_d & \mathbf{0}_d \\
\vdots & \vdots & \ddots & \vdots & \vdots & \vdots & \vdots & \ddots \quad \vdots & \vdots \\
\left(\mathbf{K}_{N^{(j)}-1,0}^{\lambda x}\right)^{(j)} & \left(\mathbf{K}_{N^{(j)}-1,1}^{\lambda x}\right)^{(j)} & \cdots & \left(\mathbf{K}_{N^{(j)}-1,N^{(j)}-1}^{\lambda x}\right)^{(j)} & \left(\mathbf{K}_{N^{(j)}-1,N^{(j)}}^{\lambda x}\right)^{(j)} & \mathbf{0}_d & \mathbf{0}_d & \cdots \left(\mathbf{K}_{N^{(j)}-1,N^{(j)}-1}^{\lambda\lambda}\right)^{(j)} & \mathbf{0}_d \\
\left(\mathbf{K}_{N^{(j)}0}^{\lambda x}\right)^{(j)} & \left(\mathbf{K}_{N^{(j)}1}^{\lambda x}\right)^{(j)} & \cdots & \left(\mathbf{K}_{N^{(j)},N^{(j)}-1}^{\lambda x}\right)^{(j)} & \left(\mathbf{K}_{N^{(j)}N^{(j)}}^{\lambda x}\right)^{(j)} & \mathbf{0}_d & \mathbf{0}_d & \cdots \quad \mathbf{0}_d & \left(\mathbf{K}_{N^{(j)}N^{(j)}}^{\lambda\lambda}\right)^{(j)}
\end{array}\right]
\end{aligned}
\tag{4.32}
$$

(4.32)

$$
\begin{aligned}
\mathbf{f}^{(j)} &= \left[\left(\mathbf{F}_1^{(j)}\right)^{\mathrm{T}}, \left(\mathbf{F}_2^{(j)}\right)^{\mathrm{T}}, \left(\mathbf{F}_3^{(j)}\right)^{\mathrm{T}}, \left(\mathbf{F}_4^{(j)}\right)^{\mathrm{T}}\right]^{\mathrm{T}} \\
&= \left[\left(\mathbf{f}_0^{x}\right)^{(j)\mathrm{T}}, \left(\mathbf{f}_1^{x}\right)^{(j)\mathrm{T}}, \ldots, \left(\mathbf{f}_{N^{(j)}-1}^{x}\right)^{(j)\mathrm{T}}, \left(\mathbf{f}_{N^{(j)}}^{x}\right)^{(j)\mathrm{T}}, \left(\mathbf{f}_0^{\lambda}\right)^{(j)\mathrm{T}}, \ldots, \left(\mathbf{f}_{N^{(j)}}^{\lambda}\right)^{(j)\mathrm{T}}\right]^{\mathrm{T}}
\end{aligned}
\tag{4.33}
$$

If we take state and costate variables at all LGL nodes to construct the following unknown variable list

$$
\mathbf{r} = \left[(\mathbf{x}_0)^{\mathrm{T}}, \left(\bar{\mathbf{x}}^{(1)}\right)^{\mathrm{T}}, \left(\bar{\lambda}^{(1)}\right)^{\mathrm{T}}; (\mathbf{x}_1)^{\mathrm{T}}, \left(\bar{\mathbf{x}}^{(2)}\right)^{\mathrm{T}}, \left(\bar{\lambda}^{(2)}\right)^{\mathrm{T}}; \cdots; (\mathbf{x}_{Q-1})^{\mathrm{T}}, \left(\bar{\mathbf{x}}^{(Q)}\right)^{\mathrm{T}}, \left(\bar{\lambda}^{(Q)}\right)^{\mathrm{T}}; (\mathbf{x}_Q)^{\mathrm{T}}\right]^{\mathrm{T}}
\tag{4.34}
$$

then the system of nonlinear algebraic equations can be expressed as

$$
\mathbf{F}(\mathbf{r}) = \mathbf{0}
\tag{4.35}
$$

where

$$
\mathbf{F} = \begin{bmatrix} \vdots \\ \left[\left(\mathbf{F}_3^{(j-1)} + \mathbf{F}_1^{(j)}\right)^{\mathrm{T}}, \left(\mathbf{F}_2^{(j)}\right)^{\mathrm{T}}, \left(\mathbf{F}_4^{(j)}\right)^{\mathrm{T}}\right]^{\mathrm{T}} \\ \left[\left(\mathbf{F}_3^{(j)} + \mathbf{F}_1^{(j+1)}\right)^{\mathrm{T}}, \left(\mathbf{F}_2^{(j+1)}\right)^{\mathrm{T}}, \left(\mathbf{F}_4^{(j+1)}\right)^{T}\right]^{\mathrm{T}} \\ \left[\left(\mathbf{F}_3^{(j+1)} + \mathbf{F}_1^{(j+2)}\right)^{\mathrm{T}}, \left(\mathbf{F}_2^{(j+2)}\right)^{\mathrm{T}}, \left(\mathbf{F}_4^{(j+2)}\right)^{T}\right]^{\mathrm{T}} \\ \vdots \end{bmatrix}^{\mathrm{T}}
\tag{4.36}
$$

and the Jacobian matrix can be expressed as

$$\mathbf{K} = \begin{bmatrix} \ddots & & & & & & & & \\ & \mathbf{K}_{33}^{(j-1)} + \mathbf{K}_{11}^{(j)} & \mathbf{K}_{12}^{(j)} & \mathbf{K}_{14}^{(j)} & \mathbf{K}_{13}^{(j)} & & & & \textit{zeros} \\ & & \mathbf{K}_{22}^{(j)} & \mathbf{K}_{24}^{(j)} & \mathbf{K}_{23}^{(j)} & & & & \\ & & & \mathbf{K}_{44}^{(j)} & \mathbf{K}_{43}^{(j)} & & & & \\ & & & & \mathbf{K}_{33}^{(j)} + \mathbf{K}_{11}^{(j+1)} & \mathbf{K}_{12}^{(j+1)} & \mathbf{K}_{14}^{(j+1)} & \mathbf{K}_{13}^{(j+1)} & \\ & & \textit{SYMM} & & & \mathbf{K}_{22}^{(j+1)} & \mathbf{K}_{24}^{(j+1)} & \mathbf{K}_{23}^{(j+1)} & \\ & & & & & & \mathbf{K}_{44}^{(j+1)} & \mathbf{K}_{43}^{(j+1)} & \\ & \textit{zeros} & & & & & & \mathbf{K}_{33}^{(j+1)} + \mathbf{K}_{11}^{(j+2)} & \\ & & & & & & & & \ddots \end{bmatrix} \tag{4.37}$$

(The upper dashed block is of size $2d\left(N^{(j)}+1\right)\times 2d\left(N^{(j)}+1\right)$; the lower dashed block is of size $2d\left(N^{(j+1)}+1\right)\times 2d\left(N^{(j+1)}+1\right)$.)

***Remark 4.1*** From the expression of matrix $\mathbf{K}$, we find it to be sparse and symmetric. And the half bandwidth of matrix $\mathbf{K}$ is $b = \max\limits_{j=1,2,\ldots Q}\left[2d\left(N^{(j)}+1\right)\right]$. Actually, since the symplectic pseudospectral method is developed based on the variational principle, $\mathbf{K}$ is naturally sparse and symmetric.

### *4.3.4 Applying the Boundary Conditions*

In this section, we discuss how to apply the boundary conditions. The system of nonlinear algebraic equations constructed in Eq. (4.35) should be modified according to the boundary conditions provided in Eqs. (4.8) or (4.9).

First, we treat the initial conditions, i.e., $\mathbf{x}_0 = \mathbf{x}_s$. Since $\mathbf{x}_0$ is a prescribed constant vector, it is removed from the unknown variable list $\mathbf{r}$. Hence, the first $d$ equations in the system of nonlinear algebraic equations are eliminated. Correspondingly, the first $d$ rows and the first $d$ columns of the Jacobian matrix are deleted.

Then, we treat the terminal conditions. And two conditions, i.e., the fixed terminal state and the free terminal state, are discussed separately.

(1) for the fixed terminal state

Under this condition, we have $\mathbf{x}_Q = \mathbf{x}_g$. Since $\mathbf{x}_Q$ is a prescribed constant vector, its treatment is similar to that for the initial conditions. $\mathbf{x}_Q$ is removed from the unknown variable list. The last $d$ equations in the system of nonlinear algebraic equations are eliminated. And the last $d$ rows and the last $d$ columns of the Jacobian matrix are deleted.

(2) for the free terminal state

As for this condition, the terminal condition is as follows:

$$\boldsymbol{\lambda}_Q = \frac{\partial \Psi(\mathbf{x}_Q)}{\partial \mathbf{x}_Q} := \mathbf{h}(\mathbf{x}_Q) \tag{4.38}$$

The last $d$ equations in the system of nonlinear algebraic equations are modified as

$$\mathbf{F}_3^{(Q)} - \mathbf{h}(\mathbf{x}_Q) = 0 \tag{4.39}$$

Hence, the last $d \times d$ elements in the Jacobian matrix should be modified correspondingly.

Now, both the initial and terminal conditions are already imposed. The system of nonlinear algebraic equations after modification is denoted as

$$\mathbf{F}_\Delta(\mathbf{r}_\Delta) = \mathbf{0} \tag{4.40}$$

where the subscript $\Delta \in \{f, g\}$ represents the type of terminal boundary condition.

### 4.3.5 *Solving the System of Nonlinear Algebraic Equations*

We solve Eq. (4.40) by the Newton–Raphson method. The results obtained in the $k$th and the $(k+1)$th iterations are denoted as $\mathbf{r}_k$ and $\mathbf{r}_{k+1}$, respectively. And the convergence criterion is set as

$$\|\mathbf{r}_{k+1} - \mathbf{r}_k\| / \|\mathbf{r}_k\| \le \varepsilon \tag{4.41}$$

where the variable $\varepsilon$ is a small quantity represents the convergent tolerance. When Eq. (4.40) is solved, control variables can be further computed according to Eq. (4.5). Thus, Problem 4 is finally solved.

## 4.4 An *hp*-Adaptive Mesh Refinement Technique

In this section, an *hp*-adaptive mesh refinement technique based on the relative curvature of state variables is developed. When the converged solution is obtained, the maximum residual error of system equations within each sub-interval is calculated first to determine whether the mesh refinement is necessary. Two methods of mesh refinement, i.e., the $h$ method and the $p$ method, are provided. In the $h$ method, a sub-interval is divided into several small sub-intervals while the approximation degree in each newly generated sub-interval keeps unchanged. In the $p$ method, the location of a sub-interval is fixed while the approximation degree increases. For a sub-interval needs the mesh refinement, the relative curvature of state variables within this sub-interval is calculated to determine whether the $h$ method or the $p$ method will be used.

### 4.4.1 Residual Error of the System Equations

Once the computational results are obtained by the SPM, according to the system Eq. (4.2) and the first-order differentiation matrix defined in Eq. (3.68), the residual error of the system equations at all LGL nodes within a sub-interval can be calculated as follows:

$$\begin{aligned} e_k^{(j)} &= \left\| \dot{\mathbf{x}}_k^{(j)} - \frac{t_j - t_{j-1}}{2}\mathbf{f}\left(\mathbf{x}_k^{(j)}, \boldsymbol{\lambda}_k^{(j)}, t_k^{(j)}\right) \right\|_\infty \\ &= \left\| \sum_{l=0}^{N^{(j)}} D_{kl}^{(j)} \mathbf{x}_l^{(j)} - \frac{t_j - t_{j-1}}{2}\mathbf{f}\left(\mathbf{x}_k^{(j)}, \boldsymbol{\lambda}_k^{(j)}, t_k^{(j)}\right) \right\|_\infty \end{aligned} \tag{4.42}$$

where $j = 1, 2, \ldots, Q$ and $k = 0, 1, \ldots, N^{(j)}$. Thus, the maximum residual error of system equation within the $j$th sub-interval is defined as follows:

$$e_{\max}^{(j)} = \max_{k=0,1,\ldots,N^{(j)}} \left(e_k^{(j)}\right) \tag{4.43}$$

We define the mesh refinement tolerance $\mu$. For the $j$th sub-interval, the mesh refinement is required if $e_{\max}^{(j)} > \mu$. It is seen that the stop criterion of the mesh refinement process is

$$e_{\max}^{(j)} \le \mu, \ \forall j \in \{1, 2, \ldots, Q\} \tag{4.44}$$

### 4.4.2 Mesh Refinement Criterion

Let $k_{q,[m]}^{(j)}$ represent the curvature of the $m$th DOF at the $q$th LGL node within the $j$th sub-interval, which is defined as

$$k_{q,[m]}^{(j)} = \left| \ddot{\mathbf{x}}_{q,[m]}^{(j)} \Big/ \left[1 + \left(\dot{\mathbf{x}}_{q,[m]}^{(j)}\right)^2\right]^{3/2} \right| \tag{4.45}$$

It is seen both the first- and the second-order differential matrices are required in Eq. (4.45). According to Eq. (3.69), the second-order differential matrix can be calculated as follow

$$\mathbf{C}^{(j)} = \mathbf{D}^{(j)}\mathbf{D}^{(j)} \tag{4.46}$$

Thus, Eq. (4.45) can be rewritten as

$$k_{q,[m]}^{(j)} = \left| \left( \sum_{l=0}^{N^{(j)}} C_{ql}^{(j)} \mathbf{x}_{l,[m]}^{(j)} \right) \Bigg/ \left[ 1 + \left( \sum_{l=0}^{N^{(j)}} D_{ql}^{(j)} \mathbf{x}_{l,[m]}^{(j)} \right)^2 \right]^{3/2} \right| \tag{4.47}$$

Then, the maximum curvature and the average curvature of the $m$th DOF within the $j$th sub-interval, which are denoted as $k_{\max,[m]}^{(j)}$ and $k_{\text{ave},[m]}^{(j)}$, respectively, are denoted as

$$k_{\max,[m]}^{(j)} = \max_{q=0,1,\ldots,N^{(j)}} \left( k_{q,[m]}^{(j)} \right) \tag{4.48}$$

$$k_{\text{ave},[m]}^{(j)} = \left( \sum\nolimits_{q=0}^{N^{(j)}} k_{q,[m]}^{(j)} \right) \Big/ \left( N^{(j)} + 1 \right) \tag{4.49}$$

Then, for the $j$th sub-interval, the relative curvature of the $m$th DOF is defined as

$$r_{[m]}^{(j)} = k_{\max,[m]}^{(j)} \Big/ k_{\text{ave},[m]}^{(j)} \tag{4.50}$$

To determine whether the $h$ method and the $p$ method should be adopted to realize the mesh refinement, a threshold $R$ is defined first. For the $j$th sub-interval, we consider the following two cases:

(1) Case I: $\exists m \in \{0, 1, \ldots, N^{(j)}\}$ such that $r_{[m]}^{(j)} > R$

Under this condition, it is thought that the state variable within the $j$th sub-interval is not smooth enough, and we implement the mesh refinement by dividing this sub-interval into several small sub-intervals as follows:

$$M^{(j)} = \text{ceil}\left[ B \cdot \lg\left( e_{\max}^{(j)} / \mu \right) \right] \tag{4.51}$$

where the parameter $B$ is a real number greater than 1. It is seen that, when $B$ and $\mu$ are given, a greater residual error of state variables will result in more sub-intervals. This is essentially an $h$ method since more accurate results are obtained by increasing the number of sub-intervals.

(2) Case II: $\forall m \in \{0, 1, \ldots, N^{(j)}\}, r_{[m]}^{(j)} \le R$

As for this case, it is thought that the state variable within the $j$th sub-interval is relatively smooth, and we implement the mesh refinement by increasing the degree of the LGL scheme used in this sub-interval as follows:

$$N_{new}^{(j)} = N^{(j)} + \text{ceil}\left[ \lg\left( e_{\max}^{(j)} / \mu \right) \right] + A \tag{4.52}$$

where the parameter $A$ is a positive integer. It is seen that, when $A$ and $\mu$ are given, a greater residual error of state variables will result in a higher approximation degree.

This is essentially a $p$ method since more accurate results are obtained by increasing the adopted degree of freedom.

### 4.4.3 The SPM Integrating with the Mesh Refinement Procedure

When integrating the SPM with the *hp*-adaptive mesh refinement procedure, an *hp*-adaptive SPM is constructed. And the *hp*-adaptive SPM mainly comes down to 8 steps:

(1) Initialize the parameters in the *hp*-adaptive mesh refinement procedure, including the maximum iteration step $iterMax$ (It is noted that $iterMax = 1$ means no mesh refinement is required), the mesh refinement tolerance $\mu$, the threshold $R$, $B$ in the $h$ method and $A$ in the $p$ method.
(2) Initialize the parameters in the SPM, including the convergent criterion $\varepsilon$.
(3) Give the initial mesh and initialize the state and costate variables at all LGL nodes.
(4) Initialize the iteration index as $iter \leftarrow 0$.
(5) Solve the problem by the SPM on the current mesh.
(6) Update the iteration index $iter \leftarrow iter + 1$
(7) If $iter = iterMax$ or Eq. (4.44) satisfies, the solving procedure ends.
(8) Calculate the new mesh parameters by the *hp*-adaptive mesh refinement technique. And go back to step (5).

## 4.5 Numerical Example

To verify the numerical precision of the developed SPM, the relative error between the computational results of state variable and its analytical solutions is denoted $\varepsilon_x$ in the following simulations.

### 4.5.1 A Problem with Analytical Solutions

Consider the following optimal control problem [1, 2]

Minimize

$$J = \frac{1}{2}\int_0^5 (x + u^2)\mathrm{d}t \tag{4.53}$$

subject to the system equation

$$\dot{x} = 2x + 2u\sqrt{x} \tag{4.54}$$

and the boundary conditions

$$x(0) = 2,\ x(5) = 1 \tag{4.55}$$

The analytical solutions to this problem are given in [1]. When using the SPM to solve this problem, the time domain is divided into $Q$ regular sub-intervals and an $N$-degree LGL scheme is used in each sub-interval. When setting $Q = 10$ and $N = 5$, converged solutions as illustrated in Figs. 4.1, 4.2 and 4.3 are obtained taking

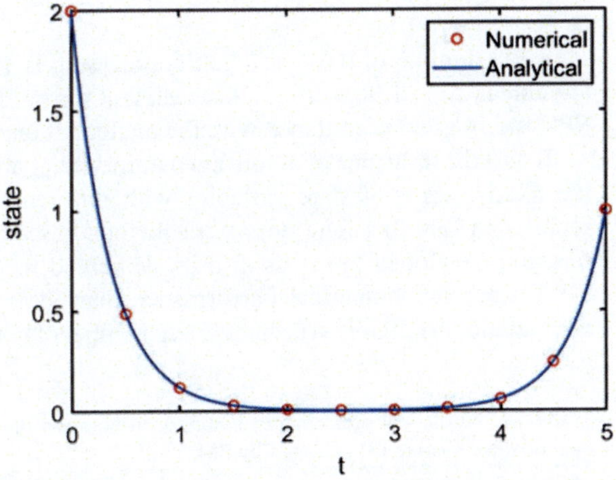


**Fig. 4.1** Numerical solutions of state

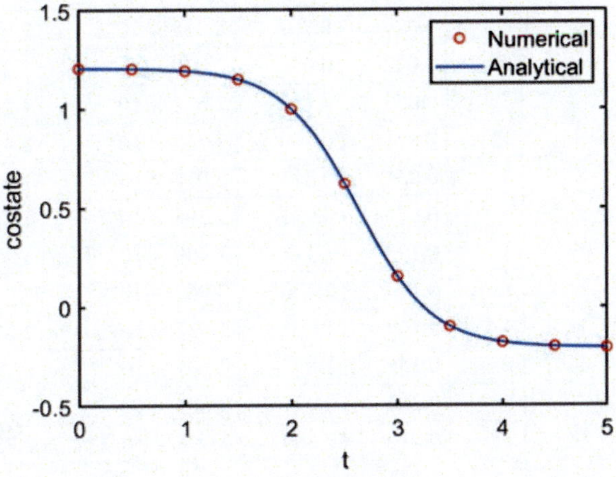


**Fig. 4.2** Numerical solutions of costate

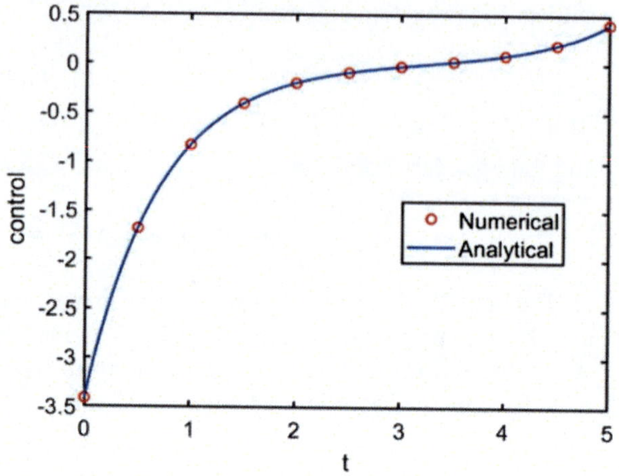


**Fig. 4.3** Numerical solutions of control

the CPU runtime of 0.062 s. The iteration step is 11 and the relative error of state variable is $\varepsilon_x = 1.90 \times 10^{-9}$. It is seen that the numerical solutions obtained by the SPM are in good accordance with the analytical ones.

To test the influence of algorithm parameters $Q$ and $N$ on the numerical precision, the relative error of state variables with different combinations of $Q$ and $N$ are reported in Table 4.1. The symplectic method based on the second kind of generating function developed in [3], is also implemented for comparison. It is seen that the SPM has better numerical performance than the symplectic method in [3], from computational efficiency, accuracy, and number of iterations. In addition, we consider

**Table 4.1** Numerical performance comparison between the developed SPM and the symplectic method developed in [3] (denoted by SM)

| $Q$ | $N$ | CPU runtime/s | | $\varepsilon_x$ | | Iteration steps | |
|---|---|---|---|---|---|---|---|
| | | SPM | SM | SPM | SM | SPM | SM |
| 4 | 4 | 0.029 | 0.787 | $8.335 \times 10^{-5}$ | $2.911 \times 10^{-3}$ | 10 | 45 |
| | 5 | 0.042 | 0.680 | $2.039 \times 10^{-6}$ | $1.239 \times 10^{-4}$ | 13 | 35 |
| | 8 | 0.059 | 0.716 | $2.522 \times 10^{-11}$ | $4.257 \times 10^{-10}$ | 12 | 29 |
| | 11 | 0.074 | 0.706 | $7.733 \times 10^{-16}$ | $8.489 \times 10^{-15}$ | 12 | 17 |
| 8 | 4 | 0.054 | 0.813 | $7.017 \times 10^{-7}$ | $1.639 \times 10^{-4}$ | 12 | 23 |
| | 5 | 0.073 | 0.979 | $1.298 \times 10^{-8}$ | $1.274 \times 10^{-6}$ | 12 | 25 |
| | 8 | 0.123 | 1.116 | $7.409 \times 10^{-16}$ | $1.209 \times 10^{-13}$ | 11 | 17 |
| | 11 | 0.199 | 2.420 | $3.902 \times 10^{-16}$ | $2.910 \times 10^{-15}$ | 11 | 31 |
| 16 | 4 | 0.109 | 1.693 | $1.731 \times 10^{-9}$ | $2.271 \times 10^{-6}$ | 11 | 21 |
| | 5 | 0.156 | 1.894 | $2.601 \times 10^{-11}$ | $6.794 \times 10^{-9}$ | 11 | 21 |
| | 8 | 0.282 | 2.921 | $3.681 \times 10^{-16}$ | $3.597 \times 10^{-15}$ | 11 | 27 |
| | 11 | 0.470 | 5.669 | $3.998 \times 10^{-16}$ | $2.738 \times 10^{-15}$ | 12 | 37 |

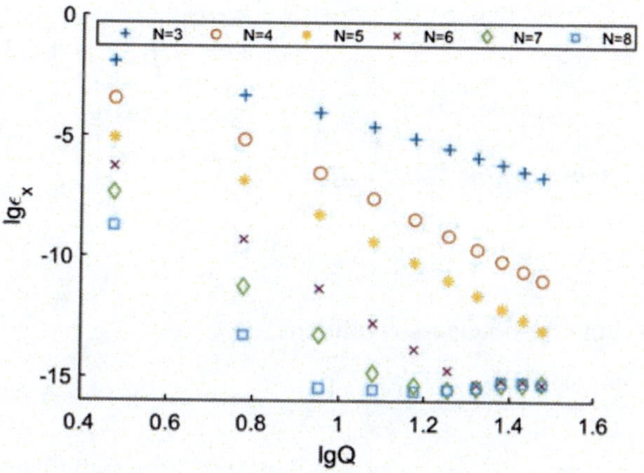


**Fig. 4.4** Numerical precision with fixed $N$ and varying $Q$

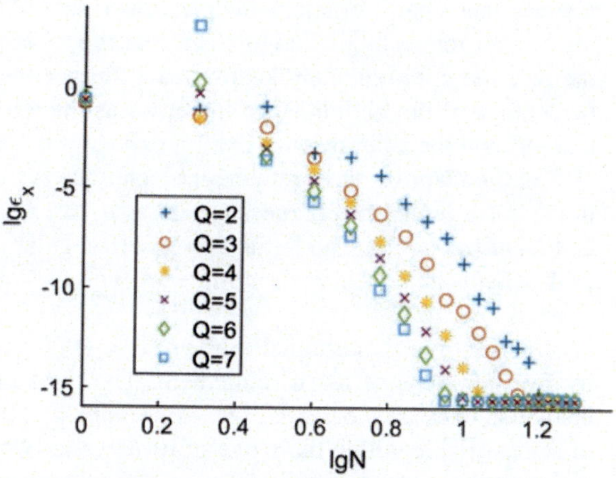


**Fig. 4.5** Numerical precision with fixed $Q$ and varying $N$

the following two cases, i.e., fixing $N$ while varying $Q$ and fixing $Q$ while varying $N$. The numerical precisions under these two conditions are plotted in Figs. 4.4 and 4.5. It is seen that the SPM could exhibit abundant convergent property. On the one hand, the SPM exhibits linear convergent property when fixing $N$ while varying $Q$; on the other hand, the SPM exhibits exponent convergent property when fixing $Q$ while varying $N$.

### 4.5.2 A Hypersensitive Problem

Consider the following hypersensitive problem

Minimize

$$J = \frac{1}{2}\int_0^{t_f} (x^2 + u^2)\mathrm{d}t \tag{4.56}$$

subject to the system equation

$$\dot{x} = -x + u \tag{4.57}$$

and the boundary conditions

$$x(0) = 1.5,\ x(t_f) = 1 \tag{4.58}$$

where $t_f$ is the given terminal time. The analytical solution to this problem is given in [4]. For a hypersensitive problem with large terminal time, its solutions exhibit a three-stage characteristic, which is composed of "take-off", "gliding", and "landing". Most interesting behaviors occur at the take-off and the landing stage, while the solution approximately keeps steady during the gliding stage. In addition, the percentage of the gliding stage increases as the terminal time becomes larger. The take-off and the landing stage exhibit exponent increasing or decreasing behavior.

The *hp*-adaptive mesh refinement technique is used in this example. Parameters in the mesh refinement process are set as $\mu = 1.0e{-6}$, $R = 1.2$, $A = 1$, $B = 1.2$, and $iterMax = 20$. The initial mesh is 10 regular sub-intervals with a two-degree LGL scheme applied in each sub-interval, and the convergent tolerance in Eq. (4.41), is set as $\varepsilon = 10^{-15}$.

When setting the terminal time as $t_f = 1000$, converged solutions are obtained by five iterations of mesh refinement. The numerical solutions together with the analytical ones are plotted in Figs. 4.6, 4.7 and 4.8. In addition, the iteration histories of state variable within the range of [0, 10] and [990, 1000] are plotted in Figs. 4.9 and 4.10, respectively.

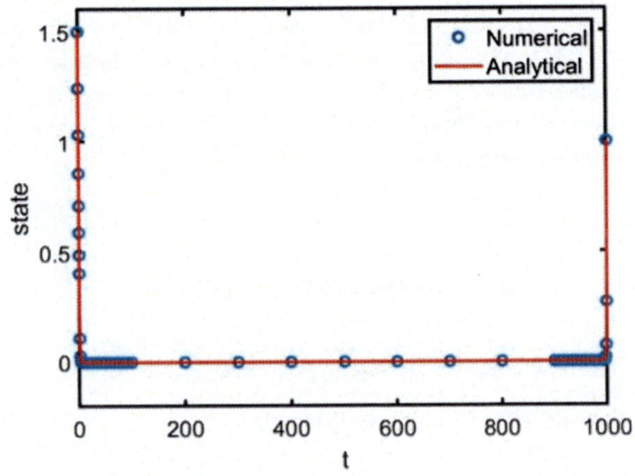


**Fig. 4.6** Numerical solutions of state

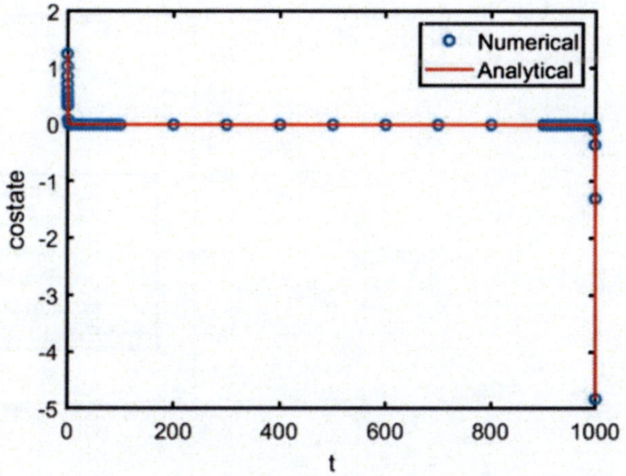


**Fig. 4.7** Numerical solutions of costate

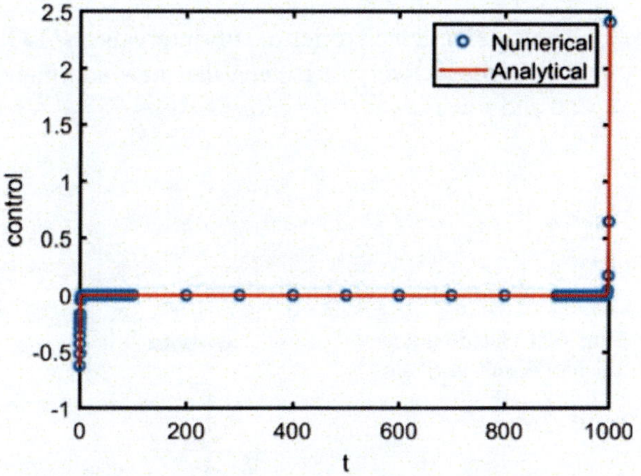


**Fig. 4.8** Numerical solutions of control

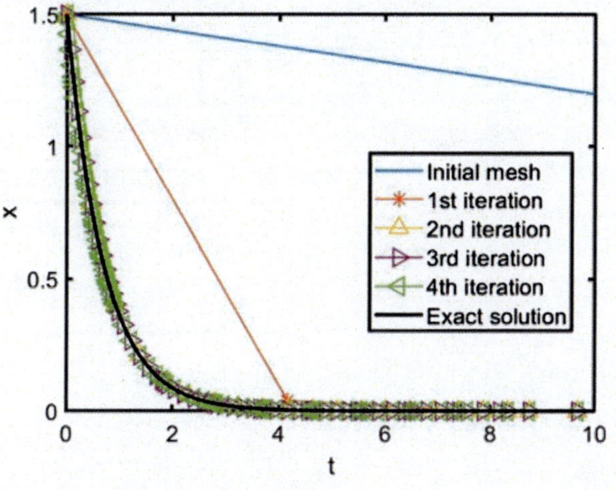


**Fig. 4.9** Iteration history of state within the range of [0, 10]

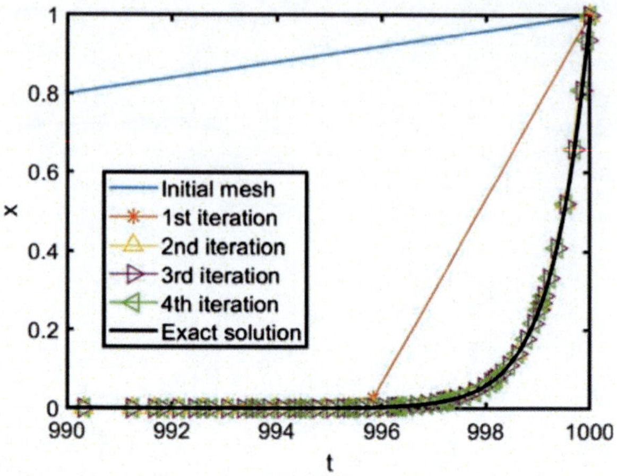


**Fig. 4.10** Iteration history of state within the range of [990, 1000]

The distribution histories of sub-intervals and LGL nodes are plotted in Figs. 4.11 and 4.12, respectively. It is seen that new sub-intervals are only generated around $t = 0$ and $t = t_f$.

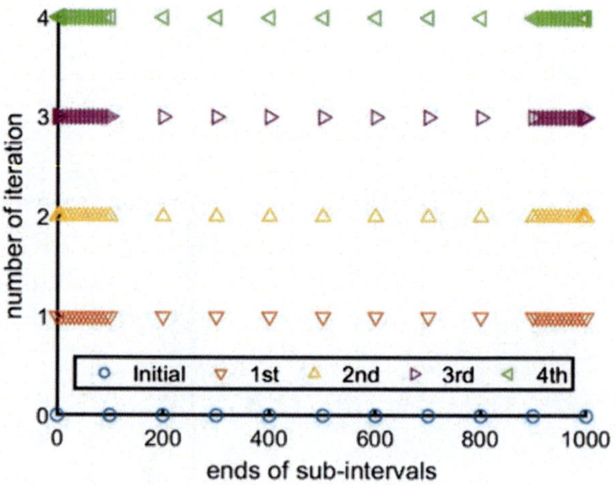


**Fig. 4.11** Distribution history of sub-intervals

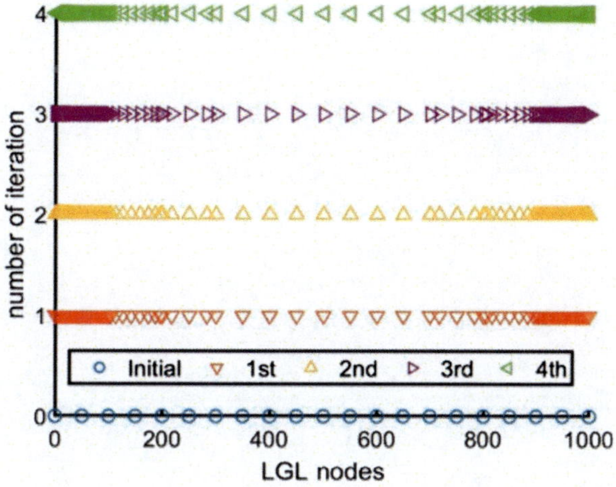


**Fig. 4.12** Distribution history of LGL nodes

# References

1. Peng H, Wang X, Li M et al (2017) An hp symplectic pseudospectral method for nonlinear optimal control. Commun Nonlinear Sci Numer Simul 42:623–644
2. Thorvaldsen TP, Huntington GT, Benson DA (2006) Direct trajectory optimization and costate estimation via an orthogonal collocation method. J Guidance Control Dyn 29(6):1435–1439
3. Peng H, Gao Q, Wu Z et al (2012) Symplectic approaches for solving two-point boundary-value problems. J Guidance Control Dyn 35(2):653–659
4. Darby CL (2011) Hp-pseudospectral method for solving continuous-time nonlinear optimal control problems [thesis]. Department of Mechanical and Aerospace Engineering, University of Florida

# Chapter 5
# SPM for Nonlinear Optimal Control Problems with Inequality Constraints

## 5.1 Introduction

Optimal control problems encountered in engineering are usually subjected to complicated constraints. According to what kinds of variables are involved, constraints can be generally categorized into three types, i.e., the pure-state constraint, the pure-control constraint, and the mixed state-control constraint. The application scenarios of such three types of constraints are introduced at first.

Pure-control constraints are the most commonly seen constraints in engineering since the inputs for the plants are usually restrained within a secure range. Hence, to avoid input saturation, box constraints are imposed, resulting in simple linear constraints. For some controlled system with multiple inputs, coupled constraints may be imposed on such inputs. For example, if the thrusts of spacecraft in three dimensions are taken as the inputs, the magnitude of the composite force is limited within a range, resulting in a nonlinear constraint [1]. Besides, for some control variable have the meaning of ratio, such as the vaccination rate in optimal control for epidemics, the control variable is naturally restrained within the range of [0, 1] [2].

One of the typical application scenarios of pure-state constraints is that inequality constraints are commonly used to describe the feasible space in trajectory planning problems [3]. Constraints on velocity are also frequently considered in case of emergency brakes. Once the velocity is taken as the state variable, constraints on velocity are pure-state ones too. In addition, for formation control problems, the relative positions of individuals are required to satisfy certain pure-state constraints [4].

As for mixed state-control constraints, they are rarely seen in mechanical systems. When solving the optimal vaccination strategy, if we take the vaccination rate among susceptible individuals $u$ and the number of individuals $S$ as control and state variables, respectively, the product of $u$ and $S$ represents the supply of vaccines at


X. Wang et al., *Symplectic Pseudospectral Methods for Optimal Control*,
Intelligent Systems, Control and Automation: Science and Engineering 97,
https://doi.org/10.1007/978-981-15-3438-6_5

each time instant. If an upper limit is imposed on the supply, a mixed state-control constraint is obtained [2].

Optimal control problems are essentially optimization problems. The Lagrange multiplier method and the penalty function method, which are the most commonly used techniques in constrained optimization problems, are also applicable to solve optimal control problems.

The Lagrange multiplier method is the most traditional method to deal with constraints in indirect methods. The constraints are introduced into the cost functional by Lagrange multipliers, then the first-order necessary conditions for optimal control problems are derived by the Pontryagin's maximum principle or the variational principle [5, 6]. In the first-order necessary conditions for constrained optimal control problems, the Hamiltonian two-point boundary value problem is coupled with other algebraic equations, enhancing the difficulty of solving procedure [7, 8]. However, constraints are strictly satisfied with the solutions obtained by the Lagrange multiplier method.

In penalty function methods, the original constrained optimization problem is transformed into an unconstrained optimization problem by adding the product of proper penalty factor and penalty function to the objective [9, 10]. It should be pointed out that penalty function methods cannot guarantee the strict satisfaction of constraints. Instead, we can only gradually decrease the violence degree of constraint by tuning the penalty factor. In addition, an unreasonable updating strategy of the penalty factor could result in non-convergence at a certain iteration step. It is seen that the penalty function method is a common technique for both the direct and indirect method since only the selection of updating strategy of penalty factors is involved.

Actually, there is no definite boundary between Lagrange multiplier methods and penalty function methods, which means that the two methods can be used together to deal with constraints. For example, to solve the loose formation control problem of constraints where the equality sphere constraints and the inequality collision-free conditions are involved simultaneously [4], the penalty function method is first adopted to deal with the equality constraints to transform the original problem with only inequality constraints. Then, a symplectic method based on the Lagrange multiplier method is developed to solve the resultant problem.

In this chapter, we focus on the optimal control problems with inequality constraints, each of the constraint can be pure-state, pure-control or mixed state-control.

## 5.2 Problem Formulation

Consider the following nonlinear optimal control problem with inequality constraints:

**Problem 5.1**
Minimize

$$J = \int_{t_s}^{t_f} L(\mathbf{x}, \mathbf{u}, t)\mathrm{d}t \tag{5.1}$$

subject to the system equation and initial conditions

$$\dot{\mathbf{x}} = \mathbf{f}(\mathbf{x}, \mathbf{u}, t),\ \mathbf{x}(t_s) = \mathbf{x}_s \tag{5.2}$$

and the inequality constraints

$$\mathbf{h}(\mathbf{x}, \mathbf{u}, t) \leq \mathbf{0} \tag{5.3}$$

where $t_s$ and $t_f$ are the fixed initial and terminal time, respectively. $\mathbf{x}(t) \in \mathbb{R}^d$ is the state variable and $\mathbf{x}_s \in \mathbb{R}^d$ represents its initial condition; $\mathbf{u}(t) \in \mathbb{R}^p$ is the control variable; $\mathbf{f} : \mathbb{R}^d \times \mathbb{R}^p \times \mathbb{R} \to \mathbb{R}^d$ is the nonlinear function to describe the system equation; $L : \mathbb{R}^d \times \mathbb{R}^p \times \mathbb{R} \to \mathbb{R}$ is the nonlinear function to describe the cost functional; $\mathbf{h} : \mathbb{R}^d \times \mathbb{R}^p \times \mathbb{R}^1 \to \mathbb{R}^q$ is a $q$-dimensional vector to describe the inequality constraints.

Two types of terminal boundary conditions, i.e., the fixed terminal state and the free terminal states are considered. As for the condition of the fixed terminal state, the prescribed terminal state is denoted as $\mathbf{x}_g \in \mathbb{R}^d$.

## 5.3 Problem Transformation

Using the successive convexification (SCvx) techniques, i.e., linearizing the system equation and the constraints and expanded the cost functional up to the second order around the nominal solutions, a series of linear-quadratic problems in the following fashion can be obtained

Minimize

$$\begin{aligned} J^{[k+1]} &= \int_{t_s}^{t_f} L^{[k+1]}(\mathbf{x}, \mathbf{u}, t)\mathrm{d}t \\ &= \int_{t_s}^{t_f} \left( \begin{array}{l} L^{[k]} + \left(\mathbf{x}^{[k+1]} - \mathbf{x}^{[k]}\right)^{\mathrm{T}} \mathbf{E}^{[k]} + \left(\mathbf{u}^{[k+1]} - \mathbf{u}^{[k]}\right)^{\mathrm{T}} \mathbf{F}^{[k]} + \frac{1}{2}\left(\mathbf{x}^{[k+1]} - \mathbf{x}^{[k]}\right)^{\mathrm{T}} \mathbf{P}^{[k]} \left(\mathbf{x}^{[k+1]} - \mathbf{x}^{[k]}\right) \\ + \left(\mathbf{u}^{[k+1]} - \mathbf{u}^{[k]}\right)^{\mathrm{T}} \mathbf{Q}^{[k]} \left(\mathbf{x}^{[k+1]} - \mathbf{x}^{[k]}\right) + \frac{1}{2}\left(\mathbf{u}^{[k+1]} - \mathbf{u}^{[k]}\right)^{\mathrm{T}} \mathbf{R}^{[k]} \left(\mathbf{u}^{[k+1]} - \mathbf{u}^{[k]}\right) \end{array} \right) \mathrm{d}t \end{aligned} \tag{5.4}$$

subject to linearized system equation and initial condition

$$\dot{\mathbf{x}}^{[k+1]} = \mathbf{A}^{[k]}\mathbf{x}^{[k+1]} + \mathbf{B}^{[k]}\mathbf{u}^{[k+1]} + \mathbf{w}^{[k]},\ \mathbf{x}^{[k+1]}(t_s) = \mathbf{x}_s \tag{5.5}$$

and linearized inequality constraints

$$\mathbf{C}^{[k]}\mathbf{x}^{[k+1]} + \mathbf{D}^{[k]}\mathbf{u}^{[k+1]} + \mathbf{v}^{[k]} \leq \mathbf{0} \tag{5.6}$$

where the notation $(\bullet)^k$ represents the result of a variable $(\bullet)$ obtained in the $k$ th iteration. And the coefficient matrices involved are express as follows:

$$L^{[k]} = L\left(\mathbf{x}^{[k]}, \mathbf{u}^{[k]}, t\right),\ \mathbf{E}^{[k]} = \left.\frac{\partial L(\mathbf{x}, \mathbf{u}, t)}{\partial \mathbf{x}}\right|_{\mathbf{x}^{[k]}, \mathbf{u}^{[k]}},\ \mathbf{F}^{[k]} = \left.\frac{\partial L(\mathbf{x}, \mathbf{u}, t)}{\partial \mathbf{u}}\right|_{\mathbf{x}^{[k]}, \mathbf{u}^{[k]}} \tag{5.7}$$

$$\mathbf{P}^{[k]} = \left.\frac{\partial^2 L(\mathbf{x}, \mathbf{u}, t)}{\partial^2 \mathbf{x}}\right|_{\mathbf{x}^{[k]}, \mathbf{u}^{[k]}},\ \mathbf{Q}^{[k]} = \left.\frac{\partial^2 L(\mathbf{x}, \mathbf{u}, t)}{\partial \mathbf{u}\partial \mathbf{x}}\right|_{\mathbf{x}^{[k]}, \mathbf{u}^{[k]}},\ \mathbf{R}^{[k]} = \left.\frac{\partial^2 L(\mathbf{x}, \mathbf{u}, t)}{\partial^2 \mathbf{u}}\right|_{\mathbf{x}^{[k]}, \mathbf{u}^{[k]}} \tag{5.8}$$

$$\mathbf{A}^{[k]} = \left.\frac{\partial \mathbf{f}(\mathbf{x}, \mathbf{u}, t)}{\partial \mathbf{x}}\right|_{\mathbf{x}^{[k]}, \mathbf{u}^{[k]}},\ \mathbf{B}^{[k]} = \left.\frac{\partial \mathbf{f}(\mathbf{x}, \mathbf{u}, t)}{\partial \mathbf{u}}\right|_{\mathbf{x}^{[k]}, \mathbf{u}^{[k]}},\ \mathbf{w}^{[k]} = \mathbf{f}^{[k]} - \mathbf{A}^{[k]}\mathbf{x}^{[k]} - \mathbf{B}^{[k]}\mathbf{u}^{[k]} \tag{5.9}$$

$$\mathbf{C}^{[k]} = \left.\frac{\partial \mathbf{h}(\mathbf{x}, \mathbf{u}, t)}{\partial \mathbf{x}}\right|_{\mathbf{x}^{[k]}, \mathbf{u}^{[k]}},\ \mathbf{D}^{[k]} = \left.\frac{\partial \mathbf{h}(\mathbf{x}, \mathbf{u}, t)}{\partial \mathbf{u}}\right|_{\mathbf{x}^{[k]}, \mathbf{u}^{[k]}},\ \mathbf{v}^{[k]} = \mathbf{h}^{[k]} - \mathbf{C}^{[k]}\mathbf{x}^{[k]} - \mathbf{D}^{[k]}\mathbf{u}^{[k]} \tag{5.10}$$

In the following iterative solving process, the results obtained in the $k$ th iteration are taken as the reference solutions for the $(k + 1)$ th iteration. And the iteration process goes on until the following convergent criterion achieves

$$\left\|\mathbf{c}^{[k+1]} - \mathbf{c}^{[k]}\right\| / \left\|\mathbf{c}^{[k+1]}\right\| \leq \varepsilon \tag{5.11}$$

where $\mathbf{c}^{[k]}$ and $\mathbf{c}^{[k+1]}$ are the solutions obtained in the $k$ th and the $(k + 1)$ th iterations, respectively. And the variable $\varepsilon$ is a small quantity that represents the convergent tolerance.

To simplify the following derivations, we consider the following problem without the iteration index.

**Problem 5.2**
Minimize

$$\begin{aligned} J &= \int_{t_s}^{t_f} \tilde{L}\mathrm{d}t \\ &= \int_{t_s}^{t_f} \begin{pmatrix} L_{\text{pre}} + \left(\mathbf{x} - \mathbf{x}_{\text{pre}}\right)^{\mathrm{T}}\mathbf{E} + \left(\mathbf{u} - \mathbf{u}_{\text{pre}}\right)^{\mathrm{T}}\mathbf{F} + \frac{1}{2}\left(\mathbf{x} - \mathbf{x}_{\text{pre}}\right)^{\mathrm{T}}\mathbf{P}\left(\mathbf{x} - \mathbf{x}_{\text{pre}}\right) \\ +\left(\mathbf{u} - \mathbf{u}_{\text{pre}}\right)^{\mathrm{T}}\mathbf{Q}\left(\mathbf{x} - \mathbf{x}_{\text{pre}}\right) + \frac{1}{2}\left(\mathbf{u} - \mathbf{u}_{\text{pre}}\right)^{\mathrm{T}}\mathbf{R}\left(\mathbf{u} - \mathbf{u}_{\text{pre}}\right) \end{pmatrix} \mathrm{d}t \end{aligned} \tag{5.12}$$

subject to linearized system equation and initial condition

$$\dot{\mathbf{x}} = \mathbf{A}\mathbf{x} + \mathbf{B}\mathbf{u} + \mathbf{w},\ \mathbf{x}(t_s) = \mathbf{x}_s \tag{5.13}$$

and linearized inequality constraints

$$\mathbf{Cx} + \mathbf{Du} + \mathbf{v} \leq \mathbf{0} \tag{5.14}$$

where the notation $(\bullet)_{\text{pre}}$ represents the result of a variable $(\bullet)$ obtained in the previous iteration.

By introducing a parametric variable $\boldsymbol{\alpha}$, the inequality constraints in Eq. (5.14), are replaced by equality ones as

$$\mathbf{Cx} + \mathbf{Du} + \mathbf{v} + \boldsymbol{\alpha} = \mathbf{0} \tag{5.15}$$

Then, according to Sect. 3.3, the first-order necessary conditions of Problem 5.2 can be derived. First, by introducing the costate variable $\boldsymbol{\lambda}$ and a parametric variable $\boldsymbol{\mu}$, the Hamiltonian of Problem 5.2 can be expressed as

$$\bar{H}(\mathbf{x}, \boldsymbol{\lambda}, \mathbf{u}, \boldsymbol{\alpha}, \boldsymbol{\mu}) = \tilde{L} + \boldsymbol{\lambda}^{\mathrm{T}}(\mathbf{Ax} + \mathbf{Bu} + \mathbf{w}) + \boldsymbol{\mu}^{\mathrm{T}}(\mathbf{Cx} + \mathbf{Du} + \mathbf{v} + \boldsymbol{\alpha}) \tag{5.16}$$

Then according to the optimality condition w.r.t. the control variable, one has

$$\frac{\partial \bar{H}(\mathbf{x}, \boldsymbol{\lambda}, \mathbf{u}, \boldsymbol{\alpha}, \boldsymbol{\mu})}{\partial \mathbf{u}} = \mathbf{0} \tag{5.17}$$

By solving Eq. (5.17), we can express the control variable as a function of state, costate and parametric variables as follows:

$$\mathbf{u} = \mathbf{u}_{\text{pre}} - \mathbf{R}^{-1}\left(\mathbf{F} + \mathbf{Q}(\mathbf{x} - \mathbf{x}_{\text{pre}}) + \mathbf{B}^{\mathrm{T}}\boldsymbol{\lambda} + \mathbf{D}^{\mathrm{T}}\boldsymbol{\mu}\right) \tag{5.18}$$

Substituting Eq. (5.18), back into Eq. (5.16), the Hamiltonian can be rewritten as follows:

$$\begin{aligned} H = {} & L_{\text{pre}} + (\mathbf{x} - \mathbf{x}_{\text{pre}})^{\mathrm{T}}\mathbf{E} + \frac{1}{2}(\mathbf{x} - \mathbf{x}_{\text{pre}})^{\mathrm{T}}\mathbf{P}(\mathbf{x} - \mathbf{x}_{\text{pre}}) \\ & - \frac{1}{2}\left(\mathbf{F}^{\mathrm{T}} + (\mathbf{x} - \mathbf{x}_{\text{pre}})^{\mathrm{T}}\mathbf{Q}^{\mathrm{T}} + \boldsymbol{\lambda}^{\mathrm{T}}\mathbf{B} + \boldsymbol{\mu}^{\mathrm{T}}\mathbf{D}\right)\mathbf{R}^{-1}\left(\mathbf{F} + \mathbf{Q}(\mathbf{x} - \mathbf{x}_{\text{pre}}) + \mathbf{B}^{\mathrm{T}}\boldsymbol{\lambda} + \mathbf{D}^{\mathrm{T}}\boldsymbol{\mu}\right) \\ & + \boldsymbol{\lambda}^{\mathrm{T}}(\mathbf{Ax} + \mathbf{w}) + \boldsymbol{\mu}^{\mathrm{T}}(\mathbf{Cx} + \mathbf{v} + \boldsymbol{\alpha}) + \left(\boldsymbol{\lambda}^{\mathrm{T}}\mathbf{B} + \boldsymbol{\mu}^{\mathrm{T}}\mathbf{D}\right)\mathbf{u}_{\text{pre}} \end{aligned} \tag{5.19}$$

And by Substituting Eq. (5.18), back into Eq. (5.15), the inequality constraint can be rewritten as follows:

$$\mathbf{Cx} + \mathbf{Du}_{\text{pre}} - \mathbf{DR}^{-1}\left(\mathbf{F} + \mathbf{Q}(\mathbf{x} - \mathbf{x}_{\text{pre}}) + \mathbf{B}^{\mathrm{T}}\boldsymbol{\lambda} + \mathbf{D}^{\mathrm{T}}\boldsymbol{\mu}\right) + \mathbf{v} + \boldsymbol{\alpha} = \mathbf{0} \tag{5.20}$$

Furthermore, the optimality conditions with respect to state and costate variables, which are also called the Hamiltonian canonical equations, can be derived as

$$\dot{\mathbf{x}} = \frac{\partial H}{\partial \boldsymbol{\lambda}} = \mathbf{A}\mathbf{x} + \mathbf{B}\left(\mathbf{u}_{\text{pre}} - \mathbf{R}^{-1}\left(\mathbf{F} + \mathbf{Q}(\mathbf{x} - \mathbf{x}_{\text{pre}}) + \mathbf{B}^{\text{T}}\boldsymbol{\lambda} + \mathbf{D}^{\text{T}}\boldsymbol{\mu}\right)\right) + \mathbf{w} \quad (5.21)$$

$$\begin{aligned}\dot{\boldsymbol{\lambda}} &= -\frac{\partial H}{\partial \mathbf{x}} \\ &= -\mathbf{E} - \mathbf{P}(\mathbf{x} - \mathbf{x}_{\text{pre}}) + \mathbf{Q}^{\text{T}}\mathbf{R}^{-1}\left(\mathbf{F} + \mathbf{Q}(\mathbf{x} - \mathbf{x}_{\text{pre}}) + \mathbf{B}^{\text{T}}\boldsymbol{\lambda} + \mathbf{D}^{\text{T}}\boldsymbol{\mu}\right) - \mathbf{A}^{\text{T}}\boldsymbol{\lambda} - \mathbf{C}^{\text{T}}\boldsymbol{\mu}\end{aligned} \quad (5.22)$$

On the other hand, according to the parametric variational principle, the parametric variables satisfy the following complementarity conditions

$$\boldsymbol{\alpha} \geq \mathbf{0},\ \boldsymbol{\mu} \geq \mathbf{0},\ \boldsymbol{\alpha}^{\text{T}}\boldsymbol{\mu} = 0 \quad (5.23)$$

It's noted that the terminal state can be either fixed or free, thus the terminal boundary condition can be derived according to Sect. 3.3. For a fixed terminal state, one has

$$\mathbf{x}(t_s) = \mathbf{x}_s,\ \mathbf{x}\left(t_f\right) = \mathbf{x}_g \quad (5.24)$$

and for the free terminal state, one has

$$\mathbf{x}(t_s) = \mathbf{x}_s,\ \boldsymbol{\lambda}\left(t_f\right) = \mathbf{0} \quad (5.25)$$

Now, the first-order necessary conditions and the boundary conditions for Problem 5.2 are already derived. It's seen that the first-order necessary conditions for this problem are the coupling of a Hamiltonian two-point boundary value problem and a linear complementarity problem.

## 5.4 Algorithm Construction

Based on the first-order necessary conditions derived in Sect. 5.3, using the second kind of generating function and the LGL local pseudospectral method, a symplectic pseudospectral method to solve Problem 5.2 is developed. The SPM is mainly divided into 6 steps:

(1) Dividing the time domain into several sub-intervals.
(2) Applying the LGL pseudospectral method to state, costate and parametric variables in each sub-interval.
(3) Applying the parametric variational principle to the second kind of generating function to form a system of linear algebraic equations.
(4) Imposing the equality and complementarity condition to every LGL nodes and to form a linear complementarity problem.

(5) Applying the boundary conditions to the system of linear algebraic equations.
(6) Establishing and solving a standard linear complementarity problem.

### 5.4.1 Time Domain Discretization

Divide the time domain $[t_s, t_f]$ into $Q$ sub-intervals, and the $j$th sub-interval is denoted as $\Gamma^{(j)} = [t_{j-1}, t_j]$, $j = 1, 2, \ldots, Q$, where $t_s = t_0 < t_1 < \ldots < t_Q = t_f$.

### 5.4.2 Applying the LGL Pseudospectral Method

To apply the LGL pseudospectral method in each sub-interval, the $j$th sub-interval is transformed into a standard time interval $[-1, 1]$ by the following linear transformation

$$\tau^{(j)} = \frac{2}{t_j - t_{j-1}}\left(t - \frac{t_j + t_{j-1}}{2}\right) \tag{5.26}$$

Thus, the second kind of generating function of the $j$th sub-interval can be written as

$$V^{(j)} = \boldsymbol{\lambda}_j^{\mathrm{T}}\mathbf{x}_j - S^{(j)} = \boldsymbol{\lambda}_j^{\mathrm{T}}\mathbf{x}_j - \int_{-1}^{1}\left(\left(\boldsymbol{\lambda}^{(j)}\right)^{\mathrm{T}}\dot{\mathbf{x}}^{(j)} - \frac{t_j - t_{j-1}}{2}H^{(j)}\right)\mathrm{d}\tau \tag{5.27}$$

In the $j$th sub-interval, state, costate and parametric variables are approximate by $N^{(j)}$-order LGL pseudospectral method as follows:

$$\mathbf{x}^{(j)}(\tau) = \sum_{l=0}^{N^{(j)}}\mathbf{x}_l^{(j)}\rho_l^{(j)}(\tau),\ \boldsymbol{\lambda}^{(j)}(\tau) = \sum_{l=0}^{N^{(j)}}\boldsymbol{\lambda}_l^{(j)}\rho_l^{(j)}(\tau) \tag{5.28}$$

$$\boldsymbol{\mu}^{(j)}(\tau) = \sum_{l=0}^{N^{(j)}}\boldsymbol{\mu}_l^{(j)}\rho_l^{(j)}(\tau),\ \boldsymbol{\alpha}^{(j)}(\tau) = \sum_{l=0}^{N(j)}\boldsymbol{\alpha}_l^{(j)}\rho_l^{(j)}(\tau) \tag{5.29}$$

As in symplectic methods based on the second kind of generating function, for any sub-interval, state at the left end and costate at the right end are taken as independent variables. Hence, for the simplicity of the following derivation, we denote the non-independent state and costate variables within the $j$th sub-interval by following notations

$$\bar{\mathbf{x}}^{(j)} = \left[\left(\mathbf{x}_1^{(j)}\right)^{\mathrm{T}}, \left(\mathbf{x}_2^{(j)}\right)^{\mathrm{T}}, \ldots, \left(\mathbf{x}_{N^{(j)}}^{(j)}\right)^{\mathrm{T}}\right]^{\mathrm{T}} \tag{5.30}$$

$$\bar{\boldsymbol{\lambda}}^{(j)} = \left[ \left( \boldsymbol{\lambda}_0^{(j)} \right)^{\mathrm{T}}, \left( \boldsymbol{\lambda}_1^{(j)} \right)^{\mathrm{T}}, \ldots, \left( \boldsymbol{\lambda}_{N^{(j)}-1}^{(j)} \right)^{\mathrm{T}} \right]^{\mathrm{T}} \tag{5.31}$$

Hence, the second kind of generating function of the $j$th sub-interval can be expressed in the following fashion by the LGL type Gauss integration

$$V^{(j)} = \left( \boldsymbol{\lambda}_{N^{(j)}}^{(j)} \right)^{\mathrm{T}} \mathbf{x}_{N^{(j)}}^{(j)} - \sum_{k=0}^{N^{(j)}} w_k^{(j)} \left( \left( \boldsymbol{\lambda}_k^{(j)} \right)^{\mathrm{T}} \left( \sum_{l=0}^{N^{(j)}} D_{kl}^{(j)} \mathbf{x}_l^{(j)} \right) - \frac{t_j - t_{j-1}}{2} H_k^{(j)} \right) \tag{5.32}$$

where $H_k^{(j)}$ represents the value of Hamiltonian at $\tau_k^{(j)}$; $w_k^{(j)}$ is the integral weight as defined in Eq. (3.57); and $\left[ D_{kl}^{(j)} \right]$ is the first-order differentiation matrix defined in Eq. (3.68).

### 5.4.3 *Applying the Variational Principle to the Second Kind of Generating Function*

According to Eq. (3.38), one has the following stationary conditions

$$\frac{\partial V^{(j)} \left( \mathbf{x}_{j-1}, \bar{\mathbf{x}}^{(j)}, \bar{\boldsymbol{\lambda}}^{(j)}, \boldsymbol{\lambda}_j \right)}{\partial \bar{\mathbf{x}}^{(j)}} = \mathbf{0} \tag{5.33}$$

$$\frac{\partial V^{(j)} \left( \mathbf{x}_{j-1}, \bar{\mathbf{x}}^{(j)}, \bar{\boldsymbol{\lambda}}^{(j)}, \boldsymbol{\lambda}_j \right)}{\partial \bar{\boldsymbol{\lambda}}^{(j)}} = \mathbf{0} \tag{5.34}$$

By summing up the action of each sub-interval, the action of the time interval $[t_s, t_f]$ is obtained. And the second kind of generating function of the time interval $[t_s, t_f]$ can be derived as follows:

$$\begin{aligned} V &= \boldsymbol{\lambda}_Q^{\mathrm{T}} \mathbf{x}_Q - \sum_{m=1}^{Q} S^{(m)} = \boldsymbol{\lambda}_Q^{\mathrm{T}} \mathbf{x}_Q - \sum_{m=1}^{Q} \left( \boldsymbol{\lambda}_m^{\mathrm{T}} \mathbf{x}_m - V^{(m)}(\mathbf{x}_{m-1}, \boldsymbol{\lambda}_m) \right) \\ &= \sum_{m=1}^{Q} V^{(m)}(\mathbf{x}_{m-1}, \boldsymbol{\lambda}_m) - \sum_{m=1}^{Q-1} \boldsymbol{\lambda}_m^{\mathrm{T}} \mathbf{x}_m \end{aligned} \tag{5.35}$$

Applying Eq. (3.38) to Eq. (5.35), one has the following stationary conditions

$$\frac{\partial V^{(j)}(\mathbf{x}_{j-1}, \boldsymbol{\lambda}_j)}{\partial \mathbf{x}_{j-1}} = \boldsymbol{\lambda}_{j-1}, \ j = 2, 3, \ldots, Q \tag{5.36}$$

$$\frac{\partial V^{(j)}(\mathbf{x}_{j-1}, \boldsymbol{\lambda}_j)}{\partial \boldsymbol{\lambda}_j} = \mathbf{x}_j, \ j = 1, 2, \ldots, Q-1 \tag{5.37}$$

Additionally, according to Eq. (3.39), the state at $t = t_s$ and costate at $t = t_f$ can be obtained as follows:

$$\boldsymbol{\lambda}_0 = \frac{\partial V}{\partial \mathbf{x}_0} = \frac{\partial V^{(1)}}{\partial \mathbf{x}_0} \tag{5.38}$$

$$\mathbf{x}_Q = \frac{\partial V}{\partial \boldsymbol{\lambda}_Q} = \frac{\partial V^{(Q)}}{\partial \boldsymbol{\lambda}_Q} \tag{5.39}$$

Thus, in any sub-interval relationships in Eqs. (5.33)–(5.34) and (5.36)–(5.39), can be put in a unified matrix form

$$\left(\mathbf{f}_0^x\right)^{(j)} = \boldsymbol{\lambda}_{j-1} \tag{5.40}$$

$$\left(\mathbf{f}_m^x\right)^{(j)} = \mathbf{0}, \ m = 1, 2, \ldots, N^{(j)} \tag{5.41}$$

$$\left(\mathbf{f}_m^\lambda\right)^{(j)} = \mathbf{0}, \ m = 0, 1, \ldots, N^{(j)} - 1 \tag{5.42}$$

$$\left(\mathbf{f}_{N^{(j)}}^x\right)^{(j)} = \mathbf{x}_j \tag{5.43}$$

where

$$\left(\mathbf{f}_m^x\right)^{(j)} = \frac{\partial V^{(j)}}{\partial \mathbf{x}_m^{(j)}} = \sum_{n=0}^{N^{(j)}} \left(\mathbf{K}_{mn}^{xx}\right)^{(j)} \mathbf{x}_n^{(j)} + \sum_{n=0}^{N^{(j)}} \left(\mathbf{K}_{mn}^{x\lambda}\right)^{(j)} \boldsymbol{\lambda}_n^{(j)} + \left(\boldsymbol{\xi}_m^x\right)^{(j)} \boldsymbol{\mu}_m^{(j)} + \left(\boldsymbol{\zeta}_m^x\right)^{(j)} \tag{5.44}$$

$$\left(\mathbf{f}_m^\lambda\right)^{(j)} = \frac{\partial V^{(j)}}{\partial \boldsymbol{\lambda}_m^{(j)}} = \sum_{n=0}^{N^{(j)}} \left(\mathbf{K}_{mn}^{\lambda x}\right)^{(j)} \mathbf{x}_n^{(j)} + \sum_{n=0}^{N^{(j)}} \left(\mathbf{K}_{mn}^{\lambda\lambda}\right)^{(j)} \boldsymbol{\lambda}_n^{(j)} + \left(\boldsymbol{\xi}_m^\lambda\right)^{(j)} \boldsymbol{\mu}_m^{(j)} + \left(\boldsymbol{\zeta}_m^\lambda\right)^{(j)} \tag{5.45}$$

And the coefficient matrices in Eqs. (5.44)–(5.45), are defined as follows:

$$\left(\mathbf{K}_{mn}^{xx}\right)^{(j)} = \frac{t_j - t_{j-1}}{2} w_m^{(j)} \left(\mathbf{P}_m^{(j)} - \left(\mathbf{Q}_m^{(j)}\right)^{\mathrm{T}} \left(\mathbf{R}_m^{(j)}\right)^{-1} \mathbf{Q}_m^{(j)}\right) \delta_m^n \tag{5.46}$$

$$\left(\mathbf{K}_{mn}^{x\lambda}\right)^{(j)} = \left(\left(\mathbf{K}_{nm}^{x\lambda}\right)^{(j)}\right)^{\mathrm{T}}$$

$$= -w_n^{(j)} D_{nm}^{(j)} \mathbf{I} + \left( \delta_m^{N^{(j)}} \mathbf{I} + \frac{t_j - t_{j-1}}{2} w_m^{(j)} \left( \left( \mathbf{A}_m^{(j)} \right)^{\mathrm{T}} - \left( \mathbf{Q}_m^{(j)} \right)^{\mathrm{T}} \left( \mathbf{R}_m^{(j)} \right)^{-1} \left( \mathbf{B}_m^{(j)} \right)^{\mathrm{T}} \right) \right) \delta_m^n \tag{5.47}$$

$$\left( \mathbf{K}_{mn}^{\lambda\lambda} \right)^{(j)} = -\frac{t_j - t_{j-1}}{2} w_m^{(j)} \mathbf{B}_m^{(j)} \left( \mathbf{R}_m^{(j)} \right)^{-1} \left( \mathbf{B}_m^{(j)} \right)^{\mathrm{T}} \delta_m^n \tag{5.48}$$

$$\left( \boldsymbol{\xi}_m^x \right)^{(j)} = \frac{t_j - t_{j-1}}{2} w_m^{(j)} \left( \left( \mathbf{C}_m^{(j)} \right)^{\mathrm{T}} - \left( \mathbf{Q}_m^{(j)} \right)^{\mathrm{T}} \left( \mathbf{R}_m^{(j)} \right)^{-1} \left( \mathbf{D}_m^{(j)} \right)^{\mathrm{T}} \right) \tag{5.49}$$

$$\left( \boldsymbol{\zeta}_m^x \right)^{(j)} = \frac{t_j - t_{j-1}}{2} w_m^{(j)} \left( \mathbf{E}_m^{(j)} - \mathbf{P}_m^{(j)} (\mathbf{x}_{\mathrm{ref}})_m^{(j)} + \left( \mathbf{Q}_m^{(j)} \right)^{\mathrm{T}} \left( \mathbf{R}_m^{(j)} \right)^{-1} \left( \mathbf{Q}_m^{(j)} (\mathbf{x}_{\mathrm{ref}})_m^{(j)} - \mathbf{F}_m^{(j)} \right) \right) \tag{5.50}$$

$$\left( \boldsymbol{\xi}_m^{\lambda} \right)^{(j)} = -\frac{t_j - t_{j-1}}{2} w_m^{(j)} \mathbf{B}_m^{(j)} \left( \mathbf{R}_m^{(j)} \right)^{-1} \left( \mathbf{D}_m^{(j)} \right)^{\mathrm{T}} \tag{5.51}$$

$$\left( \boldsymbol{\zeta}_m^{\lambda} \right)^{(j)} = \frac{t_j - t_{j-1}}{2} w_m^{(j)} \left( \mathbf{B}_m^{(j)} \left( \mathbf{R}_m^{(j)} \right)^{-1} \left( \mathbf{Q}_m^{(j)} (\mathbf{x}_{\mathrm{ref}})_m^{(j)} - \mathbf{F}_m^{(j)} \right) + \mathbf{B}_m^{(j)} (\mathbf{u}_{\mathrm{ref}})_m^{(j)} + \mathbf{w}_m^{(j)} \right) \tag{5.52}$$

For the simplicity of the following derivations, we might as well put Eqs. (5.44)–(5.45), into the following matrix form

$$\mathbf{K}^{(j)} \begin{bmatrix} \mathbf{x}_{j-1} \\ \bar{\mathbf{x}}^{(j)} \\ \bar{\boldsymbol{\lambda}}^{(j)} \\ \boldsymbol{\lambda}_j \end{bmatrix} + \begin{bmatrix} (\boldsymbol{\xi}_x)^{(j)} \\ (\boldsymbol{\xi}_\lambda)^{(j)} \end{bmatrix} \hat{\boldsymbol{\mu}}^{(j)} + \begin{bmatrix} (\boldsymbol{\zeta}_x)^{(j)} \\ (\boldsymbol{\zeta}_\lambda)^{(j)} \end{bmatrix} = \begin{bmatrix} \boldsymbol{\lambda}_{j-1} \\ \mathbf{0}_{dN^{(j)},1} \\ \mathbf{0}_{dN^{(j)},1} \\ \mathbf{x}_j \end{bmatrix}, \quad j = 1, 2, \ldots, Q \tag{5.53}$$

where

$$\mathbf{K}^{(j)} = \begin{bmatrix} \mathbf{K}_{11}^{(j)} & \mathbf{K}_{12}^{(j)} & \mathbf{K}_{13}^{(j)} & \mathbf{K}_{14}^{(j)} \\ \mathbf{K}_{21}^{(j)} & \mathbf{K}_{22}^{(j)} & \mathbf{K}_{23}^{(j)} & \mathbf{K}_{24}^{(j)} \\ \mathbf{K}_{31}^{(j)} & \mathbf{K}_{32}^{(j)} & \mathbf{K}_{33}^{(j)} & \mathbf{K}_{34}^{(j)} \\ \mathbf{K}_{41}^{(j)} & \mathbf{K}_{42}^{(j)} & \mathbf{K}_{43}^{(j)} & \mathbf{K}_{44}^{(j)} \end{bmatrix}$$

$$= \left[ \begin{array}{c|cccc|cccc|c}
\left(\mathbf{K}_{00}^{xx}\right)^{(j)} & \mathbf{0}_{d,d} & \cdots & \mathbf{0}_{d,d} & \mathbf{0}_{d,d} & \left(\mathbf{K}_{00}^{x\lambda}\right)^{(j)} & \left(\mathbf{K}_{01}^{x\lambda}\right)^{(j)} & \cdots & \left(\mathbf{K}_{0,N^{(j)}-1}^{x\lambda}\right)^{(j)} & \left(\mathbf{K}_{0N^{(j)}}^{x\lambda}\right)^{(j)} \\
\hline
\mathbf{0}_{d,d} & \left(\mathbf{K}_{11}^{xx}\right)^{(j)} & \cdots & \mathbf{0}_{d,d} & \mathbf{0}_{d,d} & \left(\mathbf{K}_{10}^{x\lambda}\right)^{(j)} & \left(\mathbf{K}_{11}^{x\lambda}\right)^{(j)} & \cdots & \left(\mathbf{K}_{1,N^{(j)}-1}^{x\lambda}\right)^{(j)} & \left(\mathbf{K}_{1N^{(j)}}^{x\lambda}\right)^{(j)} \\
\vdots & \vdots & \ddots & \vdots & \vdots & \vdots & \vdots & \ddots & \vdots & \vdots \\
\mathbf{0}_{d,d} & \mathbf{0}_{d,d} & \cdots & \left(\mathbf{K}_{N^{(j)}-1,N^{(j)}-1}^{xx}\right)^{(j)} & \mathbf{0}_{d,d} & \left(\mathbf{K}_{N^{(j)}-1,0}^{x\lambda}\right)^{(j)} & \left(\mathbf{K}_{N^{(j)}-1,1}^{x\lambda}\right)^{(j)} & \cdots & \left(\mathbf{K}_{N^{(j)}-1,N^{(j)}-1}^{x\lambda}\right)^{(j)} & \left(\mathbf{K}_{N^{(j)}-1,N^{(j)}}^{x\lambda}\right)^{(j)} \\
\mathbf{0}_{d,d} & \mathbf{0}_{d,d} & \cdots & \mathbf{0}_{d,d} & \left(\mathbf{K}_{N^{(j)}N^{(j)}}^{xx}\right)^{(j)} & \left(\mathbf{K}_{N^{(j)}0}^{x\lambda}\right)^{(j)} & \left(\mathbf{K}_{N^{(j)}1}^{x\lambda}\right)^{(j)} & \cdots & \left(\mathbf{K}_{N^{(j)},N^{(j)}-1}^{x\lambda}\right)^{(j)} & \left(\mathbf{K}_{N^{(j)}N^{(j)}}^{x\lambda}\right)^{(j)} \\
\hline
\left(\mathbf{K}_{00}^{\lambda x}\right)^{(j)} & \left(\mathbf{K}_{01}^{\lambda x}\right)^{(j)} & \cdots & \left(\mathbf{K}_{0,N^{(j)}-1}^{\lambda x}\right)^{(j)} & \left(\mathbf{K}_{0N^{(j)}}^{\lambda x}\right)^{(j)} & \left(\mathbf{K}_{00}^{\lambda\lambda}\right)^{(j)} & \mathbf{0}_{d,d} & \cdots & \mathbf{0}_{d,d} & \mathbf{0}_{d,d} \\
\left(\mathbf{K}_{10}^{\lambda x}\right)^{(j)} & \left(\mathbf{K}_{11}^{\lambda x}\right)^{(j)} & \cdots & \left(\mathbf{K}_{1,N^{(j)}-1}^{\lambda x}\right)^{(j)} & \left(\mathbf{K}_{1N^{(j)}}^{\lambda x}\right)^{(j)} & \mathbf{0}_{d,d} & \left(\mathbf{K}_{11}^{\lambda\lambda}\right)^{(j)} & \cdots & \mathbf{0}_{d,d} & \mathbf{0}_{d,d} \\
\vdots & \vdots & \ddots & \vdots & \vdots & \vdots & \vdots & \ddots & \vdots & \vdots \\
\left(\mathbf{K}_{N^{(j)}-1,0}^{\lambda x}\right)^{(j)} & \left(\mathbf{K}_{N^{(j)}-1,1}^{\lambda x}\right)^{(j)} & \cdots & \left(\mathbf{K}_{N^{(j)}-1,N^{(j)}-1}^{\lambda x}\right)^{(j)} & \left(\mathbf{K}_{N^{(j)}-1,N^{(j)}}^{\lambda x}\right)^{(j)} & \mathbf{0}_{d,d} & \mathbf{0}_{d,d} & \cdots & \left(\mathbf{K}_{N^{(j)}-1,N^{(j)}-1}^{\lambda\lambda}\right)^{(j)} & \mathbf{0}_{d,d} \\
\hline
\left(\mathbf{K}_{N^{(j)}0}^{\lambda x}\right)^{(j)} & \left(\mathbf{K}_{N^{(j)}1}^{\lambda x}\right)^{(j)} & \cdots & \left(\mathbf{K}_{N^{(j)},N^{(j)}-1}^{\lambda x}\right)^{(j)} & \left(\mathbf{K}_{N^{(j)}N^{(j)}}^{\lambda x}\right)^{(j)} & \mathbf{0}_{d,d} & \mathbf{0}_{d,d} & \cdots & \mathbf{0}_{d,d} & \left(\mathbf{K}_{N^{(j)}N^{(j)}}^{\lambda\lambda}\right)^{(j)}
\end{array} \right] \tag{5.54}$$

$$(\boldsymbol{\xi}_x)^{(j)} = \text{diag}\left((\boldsymbol{\xi}_0^x)^{(j)}, (\boldsymbol{\xi}_1^x)^{(j)}, \ldots, (\boldsymbol{\xi}_{N^{(j)}}^x)^{(j)}\right) \tag{5.55}$$

$$(\boldsymbol{\xi}_\lambda)^{(j)} = \text{diag}\left((\boldsymbol{\xi}_0^\lambda)^{(j)}, (\boldsymbol{\xi}_1^\lambda)^{(j)}, \ldots, (\boldsymbol{\xi}_{N^{(j)}}^\lambda)^{(j)}\right) \tag{5.56}$$

$$(\boldsymbol{\zeta}_x)^{(j)} = \left\{\left((\boldsymbol{\zeta}_0^x)^{(j)}\right)^{\mathrm{T}}, \left((\boldsymbol{\zeta}_1^x)^{(j)}\right)^{\mathrm{T}}, \ldots, \left((\boldsymbol{\zeta}_{N^{(j)}}^x)^{(j)}\right)^{\mathrm{T}}\right\}^{\mathrm{T}} \tag{5.57}$$

$$(\boldsymbol{\zeta}_\lambda)^{(j)} = \left\{\left((\boldsymbol{\zeta}_0^\lambda)^{(j)}\right)^{\mathrm{T}}, \left((\boldsymbol{\zeta}_1^\lambda)^{(j)}\right)^{\mathrm{T}}, \ldots, \left((\boldsymbol{\zeta}_{N^{(j)}}^\lambda)^{(j)}\right)^{\mathrm{T}}\right\}^{\mathrm{T}} \tag{5.58}$$

$$\hat{\boldsymbol{\mu}}^{(j)} = \left\{\left(\boldsymbol{\mu}_0^{(j)}\right)^{\mathrm{T}}, \left(\boldsymbol{\mu}_1^{(j)}\right)^{\mathrm{T}}, \ldots, \left(\boldsymbol{\mu}_{N^{(j)}}^{(j)}\right)^{\mathrm{T}}\right\}^{\mathrm{T}} \tag{5.59}$$

By defining the following notations

$$\boldsymbol{\xi}^{(j)} = \begin{bmatrix} (\boldsymbol{\xi}_x)^{(j)} \\ (\boldsymbol{\xi}_\lambda)^{(j)} \end{bmatrix}, \boldsymbol{\zeta}^{(j)} = \begin{bmatrix} (\boldsymbol{\zeta}_x)^{(j)} \\ (\boldsymbol{\zeta}_\lambda)^{(j)} \end{bmatrix} \tag{5.60}$$

$$\mathbf{r}^{\text{ini}} = \left\{\boldsymbol{\lambda}_0^{\mathrm{T}}, \mathbf{0}_{1,(2N^{(1)}+1)d}\right\}^{\mathrm{T}}, \mathbf{r}^{\text{ter}} = \left\{\mathbf{0}_{1,(2N^{(Q)}+1)d}, \mathbf{x}_Q^{\mathrm{T}}\right\}^{\mathrm{T}} \tag{5.61}$$

and further taking state and costate variables at all LGL nodes to construct the following unknown variable list

$$\boldsymbol{\sigma} = \left\{(\mathbf{x}_0)^{\mathrm{T}}, (\bar{\mathbf{x}}^{(1)})^{\mathrm{T}}, \left(\bar{\boldsymbol{\lambda}}^{(1)}\right)^{\mathrm{T}}, (\boldsymbol{\lambda}_1)^{\mathrm{T}}, \ldots, (\mathbf{x}_{Q-1})^{\mathrm{T}}, (\bar{\mathbf{x}}^{(Q)})^{\mathrm{T}}, \left(\bar{\boldsymbol{\lambda}}^{(Q)}\right)^{\mathrm{T}}, (\boldsymbol{\lambda}_Q)^{\mathrm{T}}\right\}^{\mathrm{T}} \tag{5.62}$$

then the relationships in Eqs. (5.33)–(5.34) and (5.36)–(5.39), can be written into the following matrix form

$$\mathbf{K}\sigma + \boldsymbol{\xi}\hat{\boldsymbol{\mu}} + \boldsymbol{\zeta} = \mathbf{r} \tag{5.63}$$

where

$$\mathbf{K} = \begin{bmatrix} \mathbf{K}_{11}^{(1)} & \mathbf{K}_{12}^{(1)} & \mathbf{K}_{13}^{(1)} & \mathbf{K}_{14}^{(1)} & & & & & & & & & & \\ \mathbf{K}_{21}^{(1)} & \mathbf{K}_{22}^{(1)} & \mathbf{K}_{23}^{(1)} & \mathbf{K}_{24}^{(1)} & & & & & & & & & & \\ \mathbf{K}_{31}^{(1)} & \mathbf{K}_{32}^{(1)} & \mathbf{K}_{33}^{(1)} & \mathbf{K}_{34}^{(1)} & & & & & & & & & & \\ \mathbf{K}_{41}^{(1)} & \mathbf{K}_{42}^{(1)} & \mathbf{K}_{43}^{(1)} & \mathbf{K}_{44}^{(1)} & -\mathbf{I} & & & & & & & & & \\ & & & -\mathbf{I} & \mathbf{K}_{11}^{(2)} & \mathbf{K}_{12}^{(2)} & \mathbf{K}_{13}^{(2)} & \mathbf{K}_{14}^{(2)} & & & & & & \\ & & & & \mathbf{K}_{21}^{(2)} & \mathbf{K}_{22}^{(2)} & \mathbf{K}_{23}^{(2)} & \mathbf{K}_{24}^{(2)} & & & & & & \\ & & & & \mathbf{K}_{31}^{(2)} & \mathbf{K}_{32}^{(2)} & \mathbf{K}_{33}^{(2)} & \mathbf{K}_{34}^{(2)} & & & & & & \\ & & & & \mathbf{K}_{41}^{(2)} & \mathbf{K}_{42}^{(2)} & \mathbf{K}_{43}^{(2)} & \mathbf{K}_{44}^{(2)} & -\mathbf{I} & & & & & \\ & & & & & & & -\mathbf{I} & \ddots & -\mathbf{I} & & & & \\ & & & & & & & & -\mathbf{I} & \mathbf{K}_{11}^{(Q)} & \mathbf{K}_{12}^{(Q)} & \mathbf{K}_{13}^{(Q)} & \mathbf{K}_{14}^{(Q)} \\ & & & & & & & & & \mathbf{K}_{21}^{(Q)} & \mathbf{K}_{22}^{(Q)} & \mathbf{K}_{23}^{(Q)} & \mathbf{K}_{24}^{(Q)} \\ & & & & & & & & & \mathbf{K}_{31}^{(Q)} & \mathbf{K}_{32}^{(Q)} & \mathbf{K}_{33}^{(Q)} & \mathbf{K}_{34}^{(Q)} \\ & & & & & & & & & \mathbf{K}_{41}^{(Q)} & \mathbf{K}_{42}^{(Q)} & \mathbf{K}_{43}^{(Q)} & \mathbf{K}_{44}^{(Q)} \end{bmatrix} \tag{5.64}$$

$$\boldsymbol{\xi} = \mathrm{diag}\left(\boldsymbol{\xi}^{(1)}, \boldsymbol{\xi}^{(2)}, \ldots, \boldsymbol{\xi}^{(Q)}\right) \tag{5.65}$$

$$\boldsymbol{\zeta} = \left\{\left(\boldsymbol{\zeta}^{(1)}\right)^{\mathrm{T}}, \left(\boldsymbol{\zeta}^{(2)}\right)^{\mathrm{T}}, \ldots, \left(\boldsymbol{\zeta}^{(Q)}\right)^{\mathrm{T}}\right\}^{\mathrm{T}} \tag{5.66}$$

$$\hat{\boldsymbol{\mu}} = \left\{\left(\hat{\boldsymbol{\mu}}^{(1)}\right)^{\mathrm{T}}, \left(\hat{\boldsymbol{\mu}}^{(2)}\right)^{\mathrm{T}}, \ldots, \left(\hat{\boldsymbol{\mu}}^{(Q)}\right)^{\mathrm{T}}\right\}^{\mathrm{T}} \tag{5.67}$$

$$\mathbf{r} = \left\{\left(\mathbf{r}^{\mathrm{ini}}\right)^{\mathrm{T}}, \mathbf{0}_{1,2d(N^{(2)}+1)}, \mathbf{0}_{1,2d(N^{(3)}+1)}, \ldots, \mathbf{0}_{1,2d(N^{(Q-1)}+1)}, \left(\mathbf{r}^{\mathrm{ter}}\right)^{\mathrm{T}}\right\}^{\mathrm{T}} \tag{5.68}$$

**Remark 5.1**
From the expression of matrix $\mathbf{K}$, we find it to be sparse and symmetric. And the half bandwidth of matrix $\mathbf{K}$ is $b = \max\limits_{j=1,2,\ldots,Q} 2d\left(N^{(j)} + 1\right)$.

### 5.4.4 Imposing the Equality Constraints and the Complementarity Conditions

As for the $j$th sub-interval, the equality constraint defined in Eq. (5.20), should be satisfied at the $N^{(j)} + 1$ LGL nodes within the sub-interval. Such relationships can be expressed as

$$\mathbf{G}_m^{(j)}\mathbf{x}_m^{(j)} - \mathbf{H}_m^{(j)}\boldsymbol{\lambda}_m^{(j)} - \mathbf{M}_m^{(j)}\boldsymbol{\mu}_m^{(j)} + \tilde{\mathbf{v}}_m^{(j)} + \boldsymbol{\alpha}_m^{(j)} = \mathbf{0},\ m = 0, 1, \ldots, N^{(j)} \tag{5.69}$$

where

$$\mathbf{G}_m^{(j)} = \mathbf{C}_m^{(j)} - \mathbf{D}_m^{(j)}\left(\mathbf{R}_m^{(j)}\right)^{-1}\left(\mathbf{Q}_m^{(j)}\right)^{\mathrm{T}} \tag{5.70}$$

$$\mathbf{H}_m^{(j)} = \mathbf{D}_m^{(j)}\left(\mathbf{R}_m^{(j)}\right)^{-1}\left(\mathbf{B}_m^{(j)}\right)^{\mathrm{T}} \tag{5.71}$$

$$\mathbf{M}_m^{(j)} = \mathbf{D}_m^{(j)}\left(\mathbf{R}_m^{(j)}\right)^{-1}\left(\mathbf{D}_m^{(j)}\right)^{\mathrm{T}} \tag{5.72}$$

$$\tilde{\mathbf{v}}_m^{(j)} = \mathbf{D}_m^{(j)}\left((\mathbf{u}_{\mathrm{ref}})_m^{(j)} - \left(\mathbf{R}_m^{(j)}\right)^{-1}\left(\mathbf{F}_m^{(j)} - \mathbf{Q}_m^{(j)}(\mathbf{x}_{\mathrm{ref}})_m^{(j)}\right)\right) + \mathbf{v}_m^{(j)} \tag{5.73}$$

The $N^{(j)}+1$ relationships in Eq. (5.69), can be expressed in the matrix form as follows:

$$\mathbf{G}^{(j)}\hat{\mathbf{x}}^{(j)} - \mathbf{H}^{(j)}\hat{\boldsymbol{\lambda}}^{(j)} - \mathbf{M}^{(j)}\hat{\boldsymbol{\mu}}^{(j)} + \hat{\mathbf{v}}^{(j)} + \hat{\boldsymbol{\alpha}}^{(j)} = \mathbf{0} \tag{5.74}$$

where

$$\mathbf{G}^{(j)} = \mathrm{diag}\left(\mathbf{G}_0^{(j)}, \mathbf{G}_1^{(j)}, \ldots, \mathbf{G}_{N^{(j)}}^{(j)}\right) \tag{5.75}$$

$$\mathbf{H}^{(j)} = \mathrm{diag}\left(\mathbf{H}_0^{(j)}, \mathbf{H}_1^{(j)}, \ldots, \mathbf{H}_{N^{(j)}}^{(j)}\right) \tag{5.76}$$

$$\mathbf{M}^{(j)} = \mathrm{diag}\left(\mathbf{M}_0^{(j)}, \mathbf{M}_1^{(j)}, \ldots, \mathbf{M}_{N^{(j)}}^{(j)}\right) \tag{5.77}$$

$$\hat{\mathbf{x}}^{(j)} = \left\{\left(\mathbf{x}_0^{(j)}\right)^{\mathrm{T}}, \left(\mathbf{x}_1^{(j)}\right)^{\mathrm{T}}, \ldots, \left(\mathbf{x}_{N^{(j)}}^{(j)}\right)^{\mathrm{T}}\right\}^{\mathrm{T}} = \left\{\mathbf{x}_{j-1}^{\mathrm{T}}, \left(\bar{\mathbf{x}}^{(j)}\right)^{\mathrm{T}}\right\}^{\mathrm{T}} \tag{5.78}$$

$$\hat{\boldsymbol{\lambda}}^{(j)} = \left\{\left(\boldsymbol{\lambda}_0^{(j)}\right)^{\mathrm{T}}, \left(\boldsymbol{\lambda}_1^{(j)}\right)^{\mathrm{T}}, \ldots, \left(\boldsymbol{\lambda}_{N^{(j)}}^{(j)}\right)^{\mathrm{T}}\right\}^{\mathrm{T}} = \left\{\left(\bar{\boldsymbol{\lambda}}^{(j)}\right)^{\mathrm{T}}, \boldsymbol{\lambda}_j^{\mathrm{T}}\right\}^{\mathrm{T}} \tag{5.79}$$

$$\hat{\mathbf{v}}^{(j)} = \left\{\left(\tilde{\mathbf{v}}_0^{(j)}\right)^{\mathrm{T}}, \left(\tilde{\mathbf{v}}_1^{(j)}\right)^{\mathrm{T}}, \ldots, \left(\tilde{\mathbf{v}}_{N^{(j)}}^{(j)}\right)^{\mathrm{T}}\right\}^{\mathrm{T}} \tag{5.80}$$

$$\hat{\boldsymbol{\alpha}}^{(j)} = \left\{\left(\boldsymbol{\alpha}_0^{(j)}\right)^{\mathrm{T}}, \left(\boldsymbol{\alpha}_1^{(j)}\right)^{\mathrm{T}}, \ldots, \left(\boldsymbol{\alpha}_{N^{(j)}}^{(j)}\right)^{\mathrm{T}}\right\}^{\mathrm{T}} \tag{5.81}$$

With the above notations, one can further impose the equality constraints at all LGL nodes within the time domain $[t_s, t_f]$ as follows:

$$\mathbf{G}\hat{\mathbf{x}} - \mathbf{H}\hat{\boldsymbol{\lambda}} - \mathbf{M}\hat{\boldsymbol{\mu}} + \hat{\mathbf{v}} + \hat{\boldsymbol{\alpha}} = \mathbf{0} \tag{5.82}$$

where

$$\mathbf{G} = \mathrm{diag}\left(\mathbf{G}^{(1)}, \mathbf{G}^{(2)}, \ldots, \mathbf{G}^{(Q)}\right) \tag{5.83}$$

$$\mathbf{H} = \mathrm{diag}\left(\mathbf{H}^{(1)}, \mathbf{H}^{(2)}, \ldots, \mathbf{H}^{(Q)}\right) \tag{5.84}$$

$$\mathbf{M} = \mathrm{diag}\left(\mathbf{M}^{(1)}, \mathbf{M}^{(2)}, \ldots, \mathbf{M}^{(Q)}\right) \tag{5.85}$$

$$\hat{\boldsymbol{\lambda}} = \left\{\left(\hat{\boldsymbol{\lambda}}^{(1)}\right)^{\mathrm{T}}, \left(\hat{\boldsymbol{\lambda}}^{(2)}\right)^{\mathrm{T}}, \ldots, \left(\hat{\boldsymbol{\lambda}}^{(Q)}\right)^{\mathrm{T}}\right\}^{\mathrm{T}} \tag{5.86}$$

$$\hat{\mathbf{v}} = \left\{\left(\hat{\mathbf{v}}^{(1)}\right)^{\mathrm{T}}, \left(\hat{\mathbf{v}}^{(2)}\right)^{\mathrm{T}}, \ldots, \left(\hat{\mathbf{v}}^{(Q)}\right)^{\mathrm{T}}\right\}^{\mathrm{T}} \tag{5.87}$$

$$\hat{\boldsymbol{\alpha}} = \left\{\left(\hat{\boldsymbol{\alpha}}^{(1)}\right)^{\mathrm{T}}, \left(\hat{\boldsymbol{\alpha}}^{(2)}\right)^{\mathrm{T}}, \ldots, \left(\hat{\boldsymbol{\alpha}}^{(Q)}\right)^{\mathrm{T}}\right\}^{\mathrm{T}} \tag{5.88}$$

Additionally, the complementarity conditions in Eq. (5.23) should be satisfied at all LGL nodes. Thus, one has

$$\hat{\boldsymbol{\alpha}} \geq \mathbf{0},\ \hat{\boldsymbol{\mu}} \geq \mathbf{0},\ \hat{\boldsymbol{\mu}}^{\mathrm{T}}\hat{\boldsymbol{\alpha}} = 0 \tag{5.89}$$

### *5.4.5 Imposing the Boundary Conditions*

In this section, we discuss how to apply the boundary conditions. The system of linear algebraic equations constructed in Eq. (5.63) should be modified according to the boundary conditions provided in Eqs. (5.24) or (5.25).

(1) for the fixed terminal state

Under this condition, Eq. (5.63) is modified as follows:

$$\mathbf{K}_g\boldsymbol{\sigma} + \boldsymbol{\xi}_g\hat{\boldsymbol{\mu}} + \boldsymbol{\zeta}_g = \mathbf{r}_g \tag{5.90}$$

where

$$\mathbf{K}_g = \begin{bmatrix} \mathbf{I} & \mathbf{0} & \mathbf{0} & \mathbf{0} & & & & & & & & & & \\ \mathbf{K}_{21}^{(1)} & \mathbf{K}_{22}^{(1)} & \mathbf{K}_{23}^{(1)} & \mathbf{K}_{24}^{(1)} & & & & & & & & & & \\ \mathbf{K}_{31}^{(1)} & \mathbf{K}_{32}^{(1)} & \mathbf{K}_{33}^{(1)} & \mathbf{K}_{34}^{(1)} & & & & & & & & & & \\ \mathbf{K}_{41}^{(1)} & \mathbf{K}_{42}^{(1)} & \mathbf{K}_{43}^{(1)} & \mathbf{K}_{44}^{(1)} & -\mathbf{I} & & & & & & & & & \\ & & & -\mathbf{I} & \mathbf{K}_{11}^{(2)} & \mathbf{K}_{12}^{(2)} & \mathbf{K}_{13}^{(2)} & \mathbf{K}_{14}^{(2)} & & & & & & \\ & & & & \mathbf{K}_{21}^{(2)} & \mathbf{K}_{22}^{(2)} & \mathbf{K}_{23}^{(2)} & \mathbf{K}_{24}^{(2)} & & & & & & \\ & & & & \mathbf{K}_{31}^{(2)} & \mathbf{K}_{32}^{(2)} & \mathbf{K}_{33}^{(2)} & \mathbf{K}_{34}^{(2)} & & & & & & \\ & & & & \mathbf{K}_{41}^{(2)} & \mathbf{K}_{42}^{(2)} & \mathbf{K}_{43}^{(2)} & \mathbf{K}_{44}^{(2)} & -\mathbf{I} & & & & & \\ & & & & & & & -\mathbf{I} & \ddots & -\mathbf{I} & & & & \\ & & & & & & & & -\mathbf{I} & \mathbf{K}_{11}^{(Q)} & \mathbf{K}_{12}^{(Q)} & \mathbf{K}_{13}^{(Q)} & \mathbf{K}_{14}^{(Q)} \\ & & & & & & & & & \mathbf{K}_{21}^{(Q)} & \mathbf{K}_{22}^{(Q)} & \mathbf{K}_{23}^{(Q)} & \mathbf{K}_{24}^{(Q)} \\ & & & & & & & & & \mathbf{K}_{31}^{(Q)} & \mathbf{K}_{32}^{(Q)} & \mathbf{K}_{33}^{(Q)} & \mathbf{K}_{34}^{(Q)} \\ & & & & & & & & & \mathbf{K}_{41}^{(Q)} & \mathbf{K}_{42}^{(Q)} & \mathbf{K}_{43}^{(Q)} & \mathbf{K}_{44}^{(Q)} \end{bmatrix} \tag{5.91}$$

$$\boldsymbol{\xi}_g = \mathrm{diag}\left(\boldsymbol{\xi}_g^{\mathrm{ini}}, \boldsymbol{\xi}^{(2)}, \boldsymbol{\xi}^{(3)} \ldots, \boldsymbol{\xi}^{(Q-1)}, \boldsymbol{\xi}^{(Q)}\right) \tag{5.92}$$

$$\boldsymbol{\zeta}_g = \left\{\left(\boldsymbol{\zeta}_g^{\mathrm{ini}}\right)^{\mathrm{T}}, \left(\boldsymbol{\zeta}^{(2)}\right)^{\mathrm{T}}, \left(\boldsymbol{\zeta}^{(3)}\right)^{\mathrm{T}} \ldots, \left(\boldsymbol{\zeta}^{(Q-1)}\right)^{\mathrm{T}}, \left(\boldsymbol{\zeta}^{(Q)}\right)^{\mathrm{T}}\right\}^{\mathrm{T}} \tag{5.93}$$

$$\mathbf{r}_g = \left\{\mathbf{x}_s^{\mathrm{T}}, \mathbf{0}_{1,2dN^{(1)}+d}, \mathbf{0}_{1,2d(N^{(2)}+1)}, \mathbf{0}_{1,2d(N^{(3)}+1)}, \ldots, \mathbf{0}_{1,2d(N^{(Q-1)}+1)}, \mathbf{0}_{1,2dN^{(Q-1)}+d}, \mathbf{x}_g^{\mathrm{T}}\right\}^{\mathrm{T}} \tag{5.94}$$

with the following notations

$$\boldsymbol{\xi}_g^{\mathrm{ini}} = \begin{bmatrix} \mathrm{diag}\left(\mathbf{0}_{d,q}, \left(\boldsymbol{\xi}_1^x\right)^{(1)}, \left(\boldsymbol{\xi}_2^x\right)^{(1)}, \ldots, \left(\boldsymbol{\xi}_{N^{(1)}}^x\right)^{(1)}\right) \\ \left(\boldsymbol{\xi}_\lambda\right)^{(1)} \end{bmatrix} \tag{5.95}$$

$$\boldsymbol{\zeta}_g^{\mathrm{ini}} = \left[\mathbf{0}_{1,d}, \left(\left(\boldsymbol{\zeta}_1^x\right)^{(1)}\right)^{\mathrm{T}}, \left(\left(\boldsymbol{\zeta}_2^x\right)^{(1)}\right)^{\mathrm{T}}, \ldots, \left(\left(\boldsymbol{\zeta}_{N^{(1)}}^x\right)^{(1)}\right)^{\mathrm{T}}, \left(\left(\boldsymbol{\zeta}_\lambda\right)^{(1)}\right)^{\mathrm{T}}\right]^{\mathrm{T}} \tag{5.96}$$

(2) for the free terminal state

As for this condition, Eq. (5.63) is modified as follows:

$$\mathbf{K}_f\boldsymbol{\sigma} + \boldsymbol{\xi}_f\hat{\boldsymbol{\mu}} + \boldsymbol{\zeta}_f = \mathbf{r}_f \tag{5.97}$$

where

$$\mathbf{K}_f = \begin{bmatrix} \mathbf{I} & \mathbf{0} & \mathbf{0} & \mathbf{0} & & & & & & & & & & \\ \mathbf{K}_{21}^{(1)} & \mathbf{K}_{22}^{(1)} & \mathbf{K}_{23}^{(1)} & \mathbf{K}_{24}^{(1)} & & & & & & & & & & \\ \mathbf{K}_{31}^{(1)} & \mathbf{K}_{32}^{(1)} & \mathbf{K}_{33}^{(1)} & \mathbf{K}_{34}^{(1)} & & & & & & & & & & \\ \mathbf{K}_{41}^{(1)} & \mathbf{K}_{42}^{(1)} & \mathbf{K}_{43}^{(1)} & \mathbf{K}_{44}^{(1)} & -\mathbf{I} & & & & & & & & & \\ & & & -\mathbf{I} & \mathbf{K}_{11}^{(2)} & \mathbf{K}_{12}^{(2)} & \mathbf{K}_{13}^{(2)} & \mathbf{K}_{14}^{(2)} & & & & & & \\ & & & & \mathbf{K}_{21}^{(2)} & \mathbf{K}_{22}^{(2)} & \mathbf{K}_{23}^{(2)} & \mathbf{K}_{24}^{(2)} & & & & & & \\ & & & & \mathbf{K}_{31}^{(2)} & \mathbf{K}_{32}^{(2)} & \mathbf{K}_{33}^{(2)} & \mathbf{K}_{34}^{(2)} & & & & & & \\ & & & & \mathbf{K}_{41}^{(2)} & \mathbf{K}_{42}^{(2)} & \mathbf{K}_{43}^{(2)} & \mathbf{K}_{44}^{(2)} & -\mathbf{I} & & & & & \\ & & & & & & & -\mathbf{I} & \ddots & -\mathbf{I} & & & & \\ & & & & & & & & -\mathbf{I} & \mathbf{K}_{11}^{(Q)} & \mathbf{K}_{12}^{(Q)} & \mathbf{K}_{13}^{(Q)} & \mathbf{K}_{14}^{(Q)} \\ & & & & & & & & & \mathbf{K}_{21}^{(Q)} & \mathbf{K}_{22}^{(Q)} & \mathbf{K}_{23}^{(Q)} & \mathbf{K}_{24}^{(Q)} \\ & & & & & & & & & \mathbf{K}_{31}^{(Q)} & \mathbf{K}_{32}^{(Q)} & \mathbf{K}_{33}^{(Q)} & \mathbf{K}_{34}^{(Q)} \\ & & & & & & & & & \mathbf{0} & \mathbf{0} & \mathbf{0} & \mathbf{I} \end{bmatrix} \tag{5.98}$$

$$\boldsymbol{\xi}_f = \mathrm{diag}\left(\boldsymbol{\xi}_f^{\mathrm{ini}}, \boldsymbol{\xi}^{(2)}, \boldsymbol{\xi}^{(3)} \ldots, \boldsymbol{\xi}^{(Q-1)}, \boldsymbol{\xi}_f^{\mathrm{ter}}\right) \tag{5.99}$$

$$\boldsymbol{\zeta}_f = \left\{\left(\boldsymbol{\zeta}_f^{\mathrm{ini}}\right)^{\mathrm{T}}, \left(\boldsymbol{\zeta}^{(2)}\right)^{\mathrm{T}}, \left(\boldsymbol{\zeta}^{(3)}\right)^{\mathrm{T}} \ldots, \left(\boldsymbol{\zeta}^{(Q-1)}\right)^{\mathrm{T}}, \left(\boldsymbol{\zeta}_f^{\mathrm{ter}}\right)^{\mathrm{T}}\right\}^{\mathrm{T}} \tag{5.100}$$

$$\mathbf{r}_f = \left\{\left(\mathbf{r}_f^{\mathrm{ini}}\right)^{\mathrm{T}}, \mathbf{0}_{1,2d\left(N^{(2)}+1\right)}, \mathbf{0}_{1,2d\left(N^{(3)}+1\right)}, \ldots, \mathbf{0}_{1,2d\left(N^{(Q-1)}+1\right)}, \mathbf{0}_{1,2d\left(N^{(Q)}+1\right)}\right\}^{\mathrm{T}} \tag{5.101}$$

with the following notations

$$\boldsymbol{\xi}_f^{\mathrm{ini}} = \boldsymbol{\xi}_g^{\mathrm{ini}} \tag{5.102}$$

$$\boldsymbol{\xi}_f^{\mathrm{ter}} = \begin{bmatrix} \left(\boldsymbol{\xi}_x\right)^{(Q)} \\ \mathrm{diag}\left(\left(\boldsymbol{\xi}_0^{\lambda}\right)^{(Q)}, \left(\boldsymbol{\xi}_1^{\lambda}\right)^{(Q)}, \ldots, \left(\boldsymbol{\xi}_{N^{(Q)}-1}^{\lambda}\right)^{(Q)}, \mathbf{0}_{d,q}\right) \end{bmatrix} \tag{5.103}$$

$$\boldsymbol{\zeta}_f^{\mathrm{ini}} = \boldsymbol{\zeta}_g^{\mathrm{ini}} \tag{5.104}$$

$$\boldsymbol{\zeta}_f^{\mathrm{ter}} = \left[\left(\left(\boldsymbol{\zeta}_x\right)^{(Q)}\right)^{\mathrm{T}}, \left(\left(\boldsymbol{\zeta}_0^{\lambda}\right)^{(Q)}\right)^{\mathrm{T}}, \left(\left(\boldsymbol{\zeta}_1^{\lambda}\right)^{(Q)}\right)^{\mathrm{T}}, \ldots, \left(\left(\boldsymbol{\zeta}_{N^{(Q)}-1}^{\lambda}\right)^{(Q)}\right)^{\mathrm{T}}, \mathbf{0}_{1,d}\right] \tag{5.105}$$

For the simplicity of the following derivations, Eqs. (5.90) and (5.97) are written into a uniform fashion as follows:

$$\mathbf{K}_\Delta \boldsymbol{\sigma} + \boldsymbol{\xi}_\Delta \hat{\boldsymbol{\mu}} + \boldsymbol{\zeta}_\Delta = \mathbf{r}_\Delta \tag{5.106}$$

where the subscript $\Delta \in \{f, g\}$ represents the type of terminal boundary condition.

It's seen that $\hat{\mathbf{x}}$ and $\hat{\boldsymbol{\lambda}}$ involves in the equality constraint (5.82), such two variables include information on state and costate variables at all LGL nodes. And $\boldsymbol{\sigma}$ also includes information on state and costate variables at all LGL nodes. Thus, one can view $\boldsymbol{\sigma}$ as the rearrangement of $\hat{\mathbf{x}}$ and $\hat{\boldsymbol{\lambda}}$. Actually, Eq. (5.106), together with Eqs. (5.82) and (5.89), comprise a mixed linear complementarity problem.

According to Eq. (5.106), one has

$$\boldsymbol{\sigma} = -\mathbf{K}_\Delta^{-1}\boldsymbol{\xi}_\Delta \hat{\boldsymbol{\mu}} + \mathbf{K}_\Delta^{-1}(\mathbf{r}_\Delta - \boldsymbol{\zeta}_\Delta) \text{ or } \boldsymbol{\sigma} = \boldsymbol{\Omega}\hat{\boldsymbol{\mu}} + \boldsymbol{\varphi} \tag{5.107}$$

where

$$\boldsymbol{\Omega} = -\mathbf{K}_\Delta^{-1}\boldsymbol{\xi}_\Delta \tag{5.108}$$

$$\boldsymbol{\varphi} = \mathbf{K}_\Delta^{-1}(\mathbf{r}_\Delta - \boldsymbol{\zeta}_\Delta) \tag{5.109}$$

Thus, $\hat{\mathbf{x}}$ and $\hat{\boldsymbol{\lambda}}$ can be expressed as linear functions of $\hat{\boldsymbol{\mu}}$ as follows:

$$\hat{\mathbf{x}} = \boldsymbol{\Omega}_x \hat{\boldsymbol{\mu}} + \boldsymbol{\varphi}_x \tag{5.110}$$

$$\hat{\boldsymbol{\lambda}} = \boldsymbol{\Omega}_\lambda \hat{\boldsymbol{\mu}} + \boldsymbol{\varphi}_\lambda \tag{5.111}$$

where the expressions of $\boldsymbol{\Omega}_x$, $\boldsymbol{\varphi}_x$, $\boldsymbol{\Omega}_\lambda$, and $\boldsymbol{\varphi}_\lambda$ are as follows:

$$(.)_x = \left\{ \begin{array}{c} (.)\left[2d\sum_{k=1}^{0}\left(N^{(k)}+1\right)+1 : 2d\sum_{k=1}^{0}\left(N^{(k)}+1\right)+d\left(N^{(1)}+1\right)\right] \\ (.)\left[2d\sum_{k=1}^{1}\left(N^{(k)}+1\right)+1 : 2d\sum_{k=1}^{1}\left(N^{(k)}+1\right)+d\left(N^{(2)}+1\right)\right] \\ \vdots \\ (.)\left[2d\sum_{k=1}^{Q-1}\left(N^{(k)}+1\right)+1 : 2d\sum_{k=1}^{Q-1}\left(N^{(k)}+1\right)+d\left(N^{(Q)}+1\right)\right] \end{array} \right\} \tag{5.112}$$

$$
(.)_\lambda = \left\{ \begin{array}{l}
(.)\left[2d\sum_{k=1}^{0}\left(N^{(k)}+1\right)+d\left(N^{(1)}+1\right)+1:2d\sum_{k=1}^{1}\left(N^{(k)}+1\right)\right] \\
(.)\left[2d\sum_{k=1}^{1}\left(N^{(k)}+1\right)+d\left(N^{(1)}+1\right)+1:2d\sum_{k=1}^{2}\left(N^{(k)}+1\right)\right] \\
\vdots \\
(.)\left[2d\sum_{k=1}^{Q-1}\left(N^{(k)}+1\right)+d\left(N^{(1)}+1\right)+1:2d\sum_{k=1}^{Q}\left(N^{(k)}+1\right)\right]
\end{array} \right\} \tag{5.113}
$$

where the variable (.) can refer to either $\boldsymbol{\Omega}$ or $\boldsymbol{\varphi}$.

### 5.4.6 Solving the Standard Linear Complementarity Problem

Substituting Eqs. (5.110)–(5.111) into the equality constraints (5.82), one has the following relationship between the parametric variable $\hat{\boldsymbol{\alpha}}$ and the parametric variable $\hat{\boldsymbol{\mu}}$

$$
Y\hat{\boldsymbol{\mu}} + \mathbf{q} = \hat{\boldsymbol{\alpha}} \tag{5.114}
$$

where

$$
\mathbf{Y} = \mathbf{M} - \mathbf{G}\boldsymbol{\Omega}_x + \mathbf{H}\boldsymbol{\Omega}_x \tag{5.115}
$$

$$
\mathbf{q} = \mathbf{H}\boldsymbol{\varphi}_\lambda - \mathbf{G}\boldsymbol{\varphi}_\lambda - \hat{\mathbf{v}} \tag{5.116}
$$

Thus, Eq. (5.114), together with Eq. (5.89), comprise a standard linear complementarity problem w.r.t. the parametric variable $\hat{\boldsymbol{\mu}}$ as follows:

$$
\left\{ \begin{array}{l}
\hat{\boldsymbol{\mu}} \geq 0 \\
\mathbf{Y}\hat{\boldsymbol{\mu}} + \mathbf{q} \geq 0 \\
\hat{\boldsymbol{\mu}}^{\mathrm{T}}\left(\mathbf{Y}\hat{\boldsymbol{\mu}} + \mathbf{q}\right) = 0
\end{array} \right. \tag{5.117}
$$

By applying the Lemke's method to Eq. (5.117), $\hat{\boldsymbol{\mu}}$ is solved. Then according to Eqs. (5.110)–(5.111), $\hat{\mathbf{x}}$ and $\hat{\boldsymbol{\lambda}}$ are solved. And control variable at all LGL nodes can be solved according to Eq. (5.18). Thus, Problem 5.2 is finally solved.

## 5.5 Treatment of Unspecific Terminal Time

In the previous sections of this chapter, we simply assume that the terminal time $t_f$ is fixed. However, this restriction is rarely satisfied with optimal control problems encountered in practical engineering. For most of the control problems, it is impossible to determine the terminal time in advance. Hence, developing SPM with unspecific terminal time is necessary.

In most direct methods, the terminal time variable, as well as state and control variables are taken as an unknown variable. It greatly increases the nonlinearity of the resulted NLP. In this section, according to the optimality condition with respect to the terminal time, we develop an iterative manner based on the secant method to solve problems with unspecific time.

### *5.5.1 Formulation of Problems with Unspecific Terminal Time*

Consider the following nonlinear optimal control problem with unspecific terminal time

**Problem 5.3**

$$\begin{cases} \min\ J = \Psi\big(t_f, \mathbf{x}(t_f)\big) + \int_{t_s}^{t_f} L(\mathbf{x}, \mathbf{u}, t)\mathrm{d}t \\ \text{s.t.} \\ \dot{\mathbf{x}} = \mathbf{f}(\mathbf{x}, \mathbf{u}, t),\ \mathbf{x}(t_s) = \mathbf{x}_s \\ \mathbf{h}(\mathbf{x}, \mathbf{u}, t) \le \mathbf{0} \end{cases} \tag{5.118}$$

The Hamiltonian of Problem 5.3 is defined as

$$\bar{H} = L(\mathbf{x}, \mathbf{u}, t) + \boldsymbol{\lambda}^{\mathrm{T}}\mathbf{f}(\mathbf{x}, \mathbf{u}, t) + \boldsymbol{\mu}^{\mathrm{T}}\mathbf{h}(\mathbf{x}, \mathbf{u}, t) \tag{5.119}$$

According to the optimality condition with respect to control variable, one can rewrite the Hamiltonian in Eq. (5.119), as a function of state, costate, parametric, and time variables, i.e., $H(\mathbf{x}, \boldsymbol{\lambda}, \boldsymbol{\mu}, \boldsymbol{\alpha}, t)$. Compared to Problem 5.1, for Problem 5.3, an extra optimality condition should be supplemented as follows [5]:

$$0 = \partial\Psi / \partial t_{\mathrm{f}} + H|_{t=t_{\mathrm{f}}} \tag{5.120}$$

### 5.5.2 A Secant Method to Determine the Optimal Terminal Time

Based on Eq. (5.120), one can construct a series of sub-problems with different terminal time, i.e., Problem $\mathrm{P}^{[k]}$ in the following fashion

$$\text{Problem P}^{[k]}: \begin{cases} \min J = \Psi\left(t_\mathrm{f}^{[k]}, \mathbf{x}(t_f)\right) + \int_{t_\mathrm{s}}^{t_\mathrm{f}^{[k]}} L(\mathbf{x}, \mathbf{u}, t)\mathrm{d}t \\ \text{s.t.} \\ \dot{\mathbf{x}} = \mathbf{f}(\mathbf{x}, \mathbf{u}, t), \mathbf{x}(t_\mathrm{s}) = \mathbf{x}_\mathrm{s} \\ \mathbf{h}(\mathbf{x}, \mathbf{u}, t) \leq \mathbf{0} \end{cases} \tag{5.121}$$

The Hamiltonian of Problem $\mathrm{P}^{[k]}$ is denoted as $H^{[k]}$. We might as well define the following notation

$$F^{[k]} = \partial\Psi \Big/ \partial t_\mathrm{f}^{[k]} + H^{[k]}\Big|_{t=t_\mathrm{f}^{[k]}} \tag{5.122}$$

According to Eq. (5.120), for a sub-problem with the optimal terminal time, it should satisfy that $F^{[k]} = 0$. Hence, an iterative method based on the secant method can be developed to determine the optimal terminal time (as illustrated in Fig. 5.1). The method consists of the following six steps:

Step 1: Select two initial guesses of the terminal times (i.e., $t_\mathrm{f}^{[1]}$ and $t_\mathrm{f}^{[2]}$), and construct corresponding sub-problems (i.e., Problem $\mathrm{P}^{[1]}$ and Problem $\mathrm{P}^{[2]}$). Solve Problem $\mathrm{P}^{[1]}$ and Problem $\mathrm{P}^{[2]}$, and calculate $F^{[1]}$ and $F^{[2]}$.

Step 2: Set the convergent tolerance on optimality condition $\varepsilon'$, and the convergent tolerance on terminal time variation $\tau$. Set the iteration index $k = 2$.

Step 3: Update the iteration index $k = k + 1$.

Step 4: According to Fig. 5.1, determine the terminal time for Problem $\mathrm{P}^{[k]}$ as

$$t_\mathrm{f}^{[k]} = \frac{F^{[k-1]}t_\mathrm{f}^{[k-2]} - F^{[k-2]}t_\mathrm{f}^{[k-1]}}{F^{[k-1]} - F^{[k-2]}} \tag{5.123}$$

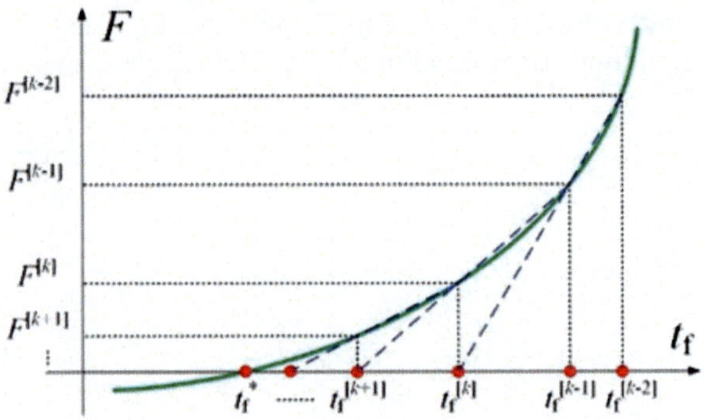


**Fig. 5.1** An illustration of using the secant method to determine the optimal terminal time

Step 5: Taking the solutions of Problem $P^{[k-1]}$ as initial guesses to solve Problem $P^{[k]}$ by the SPM provided in Sects. 5.3 and 5.4. Then calculate $F^{[k]}$.

Step 6: If $\left|F^{[k]}\right| \leq \varepsilon'$ or $\left|t_f^{[k]} - t_f^{[k-1]}\right| \leq \tau$ establishes, convergence is considered to be achieved, i.e., $t_f^{[k]}$ and solutions for Problem $P^{[k]}$ is the optimal solutions for the original problem; otherwise, go back to Step 3.

## 5.6 Numerical Example

### 5.6.1 The Breakwell Problem

Consider the classical Breakwell problem where pure-state constraints are considered. The analytical solutions of this problem can be derived, and thus the problem is often taken as the benchmark for numerical methods to solve nonlinear constrained optimal control problems. The problem aims at minimizing the following cost functional

$$J = \frac{1}{2}\int_0^1 u^2 \mathrm{d}t \tag{5.124}$$

subject to the linear system equation

$$\begin{cases} \dot{x}_1 = x_2 \\ \dot{x}_2 = u \end{cases} \tag{5.125}$$

boundary conditions

$$\begin{cases} x_1(0) = x_1(1) = 0 \\ x_2(0) = -x_2(1) = 1 \end{cases} \tag{5.126}$$

and the pure-state constraint

$$x_1 \leq l \tag{5.127}$$

where the parameter $l$ varies in the range of $\left[\frac{1}{12}, \frac{1}{3}\right]$, and the characterization of the optimal solutions varies according to the value of $l$ [5]. Denote the optimal solutions of state $x_1$, state $x_2$, costate $\lambda_1$, costate $\lambda_2$, control $u$ and cost functional $J$ as $x_1^*$, $x_2^*$, $\lambda_1^*$, $\lambda_2^*$, $u^*$, and $J^*$, respectively. Consider the following three conditions

(1) when $\frac{1}{4} < l \leq \frac{1}{3}$, one has

$$x_1^*(t)=t(1-t),\ x_2^*(t) = 1-2t,\ \lambda_1^*(t) = \lambda_2^*(t) = 0,\ u^*(t) = -2,\ J^* = 2 \tag{5.128}$$

(2) (2) when $\frac{1}{6} \leq l \leq \frac{1}{4}$, one has

$$x_1^*(t) = \begin{cases} t - 4(1-3l)t^2 + 4(1-4l)t^3, & 0 \leq t \leq \frac{1}{2} \\ 1 - t - 4(1-3l)(1-t)^2 + 4(1-4l)(1-t)^3, & \frac{1}{2} < t \leq 1 \end{cases} \tag{5.129}$$

$$x_2^*(t) = \begin{cases} 1 - 8(1-3l)t + 12(1-4l)t^2, & 0 \leq t \leq \frac{1}{2} \\ -1 + 8(1-3l)(1-t) - 12(1-4l)(1-t)^2, & \frac{1}{2} < t \leq 1 \end{cases} \tag{5.130}$$

$$\lambda_1^*(t)=\begin{cases} 24(1-4l), & 0 \leq t \leq \frac{1}{2} \\ -24(1-4l), & \frac{1}{2} < t \leq 1 \end{cases} \tag{5.131}$$

$$\begin{aligned}\lambda_2^*(t) &= -u^*(t) \\ &= \begin{cases} 8(1-3l) - 24(1-4l)t, & 0 \leq t \leq \frac{1}{2} \\ 8(1-3l) - 24(1-4l)(1-t)8(1-3l) - 24(1-4l)(1-t), & \frac{1}{2} < t \leq 1 \end{cases}\end{aligned} \tag{5.132}$$

$$J^* = 2+6(1-4l)^2 \tag{5.133}$$

(3) when $\frac{1}{12} \leq l < \frac{1}{6}$, one has

$$x_1^*(t) = \begin{cases} l\left[1 - \left(1 - \frac{t}{3l}\right)^3\right]l\left[1 - \left(1 - \frac{t}{3l}\right)^3\right], & 0 \leq t < 3l \\ l, & 3l \leq t \leq 1-3l \\ l\left[1 - \left(1 - \frac{1-t}{3l}\right)^3\right], & 1-3l < t \leq 1 \end{cases} \tag{5.134}$$

$$x_2^*(t) = \begin{cases} \left(1 - \frac{t}{3l}\right)^2, & 0 \leq t < 3l \\ 0 & 3l \leq t \leq 1-3l \\ -\left(1 - \frac{1-t}{3l}\right)^2, -\left(1 - \frac{1-t}{3l}\right)^2, & 1-3l < t \leq 1 \end{cases} \tag{5.135}$$

$$\lambda_1^*(t) = \begin{cases} 2/9l^2, & 0 \leq t < 3l \\ 0, & 3l \leq t \leq 1-3l \\ -2/9l^2, & 1-3l < t \leq 1 \end{cases} \tag{5.136}$$

$$\lambda_2^*(t) = -u^*(t) = \begin{cases} \frac{2}{3l}\left(1 - \frac{t}{3l}\right), & 0 \leq t < 3l \\ 0, & 3l \leq t \leq 1-3l \\ \frac{2}{3l}\left(1 - \frac{1-t}{3l}\right), & 1-3l < t \leq 1 \end{cases} \tag{5.137}$$

$$J^* = 4/9l \tag{5.138}$$

In the following simulations, the approximation degree of the LGL pseudospectral method applied in each sub-interval is set to an identical value, i.e., $N$. Consider the convergent criterion Eq. (5.11), we set $\mathbf{c} = \hat{\mathbf{x}}$ and $\varepsilon = 10^{-10}$.

When setting parameters in the SPM as $Q = 10$ and $N = 5$ with a regular mesh, numerical solutions of state, costate and control variables with different values of $l$ is provided in Figs. 5.2, 5.3, 5.4, 5.5, and 5.6. Additionally, the numerical solution of the cost functional together with the analytical one is plotted in Fig. 5.7. From Fig. 5.7, it is seen that the numerical solution is in good agreement with the analytical one. Noticing that the system equation and the inequality constraints are linear while the cost functional is quadratic, the Breakwell problem itself is a linear-quadratic problem. Hence, it's no need to incorporating the SCvx technique and the converged solutions are obtained by only one iteration.

From Eq. (5.136) or Fig. 5.4, it is seen that there are discontinuities in the costate $\lambda_1$ when $l \leq \frac{1}{6}$. In the simulations above we set $Q = 10$, thus there is always a

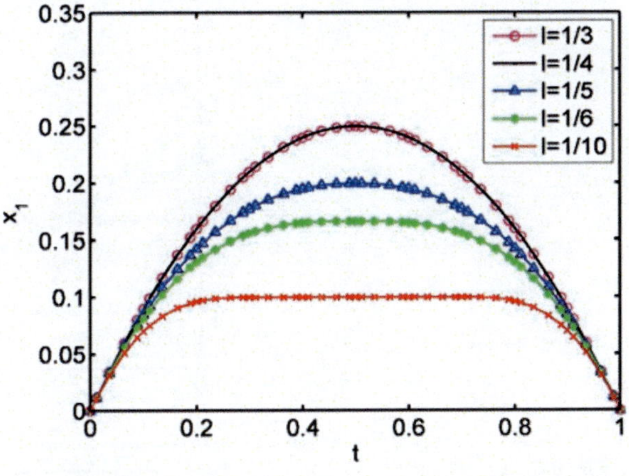


**Fig. 5.2** Numerical solutions of $x_1$

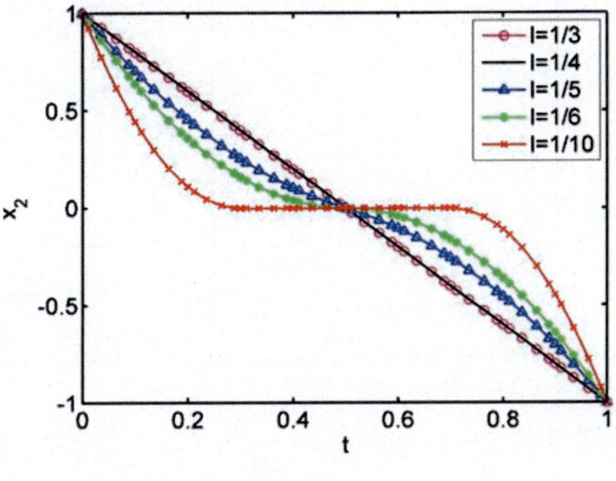


**Fig. 5.3** Numerical solutions of $x_2$

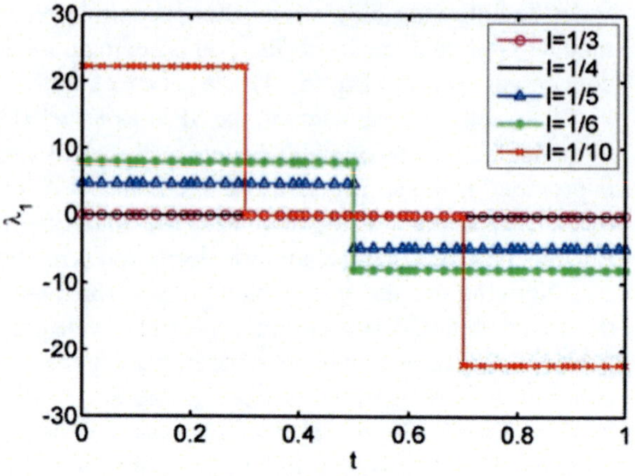


**Fig. 5.4** Numerical solutions of $\lambda_1$

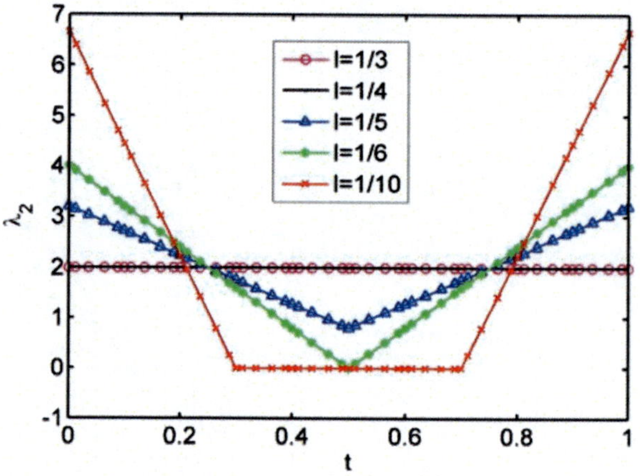


**Fig. 5.5** Numerical solutions of $\lambda_2$

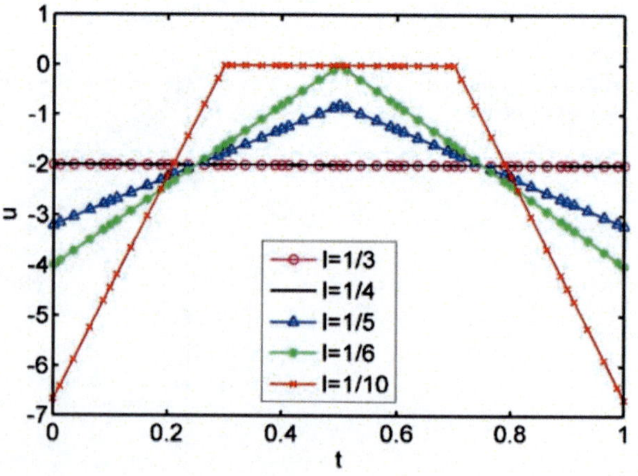


**Fig. 5.6** Numerical solutions of $u$

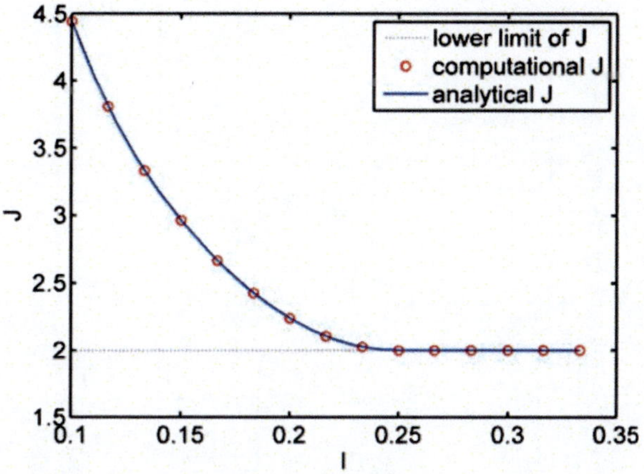


**Fig. 5.7** Numerical and analytical solution of the cost functional

sub-interval end distributed at the discontinuity of $\lambda_1$. In the following simulations, we set $l = 1/12$ and discuss the influence of sub-interval distribution on numerical precision. When setting $l = 1/12$, the optimal solution of $\lambda_1$ exhibits a three-stage characteristic, and the three stages are $T_1 = [0, 3l]$, $T_2 = [3l, 1 - 3l]$ and $T_3 = [1 - 3l, 1]$, respectively. To quantify the influence of sub-interval distribution, we denote the numerical relative errors of $x_1$, $x_2$, $\lambda_1$, and $\lambda_2$ as $\varepsilon_{x1}$, $\varepsilon_{x2}$, $\varepsilon_{\lambda 1}$, and $\varepsilon_{\lambda 2}$, respectively.

(1) Condition 1: There is no sub-interval end distributed at the discontinuity of costate

Under this condition, we test two cases, i.e., *Case* 1: fixing $N$ while tuning $Q$ and *Case* 2: fixing $Q$ while tuning $N$. The precision of state and costate variables in such two cases are given in Fig. 5.8 and Fig. 5.9, respectively.

In Fig. 5.8, we set the parameter $N \in \{k\}_{k=1}^{5}$ while $Q \in \{4k - 1\}_{k=1}^{9}$. The time domain is divided into regular sub-intervals. However, it should be noted that no matter how many sub-interval intervals are divided, the costate $\lambda_1$ cannot converge if we set $N$ equals 1 or 2. Simulations show that though the relative errors of such four variables decrease when increasing the number of sub-intervals, the convergent velocity of the costate $\lambda_1$ is slower than the other three variables. Additionally, it is seen that the SPM exhibits the property of linear convergence when fixing $N$ while tuning $Q$.

In Fig. 5.9, we set the parameter $N \in \{k\}_{k=3}^{20}$ while $Q = 5$ (as stated previously, the convergence of the costate $\lambda_1$ cannot be achieved when $N < 3$). From the overall trend, it is seen that the relative errors of such four variables decrease when increasing the approximation degree. However, the convergent velocity of the costate $\lambda_1$ is still slower than the other three variables. Furthermore, similar to the traditional direct pseudospectral method, the property of exponential convergence is lost.

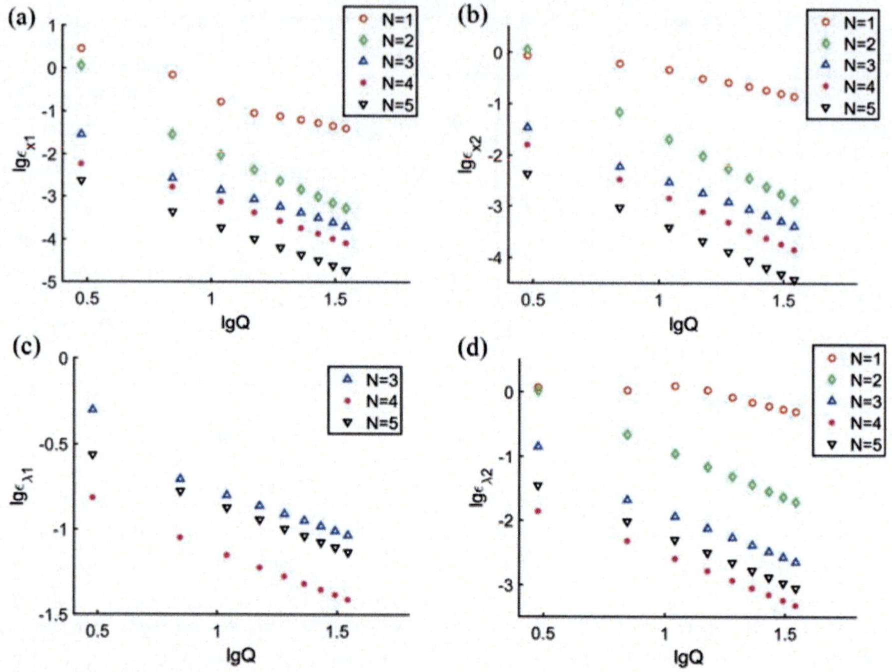


**Fig. 5.8** Numerical precision with fixed $N$ and varying $Q$

This problem is also studied by the adaptive $p$ pseudospectral method and the adaptive $h$-2 pseudospectral method in [11] and the adaptive $hp$-3 pseudospectral method in [12]. The precisions of costate variables obtained in different numerical methods are reported in Table 5.1, where the notation "×" means cannot converge. It is seen that the SPM method has better precision than other numerical methods, especially of the costate $\lambda_1$.

(2) Condition 2: There are sub-interval ends distributed at the discontinuity of costate

Under this condition, each of the three stages, i.e., $T_1 = [0, 3l]$, $T_2 = [3l, 1 - 3l]$, and $T_3 = [1 - 3l, 1]$ are divided into $\tilde{Q}$ regular sub-intervals. Thus, the time domain $[0, 1]$ is divided into $Q = 3\tilde{Q}$ sub-intervals. If $N$ varies from $\{3, 4, 5\}$ meanwhile $\tilde{Q}$ varies from $\{2, 3, 4, 5, 6\}$, the numerical precision of state and costate variables are plotted in Fig. 5.10. It is seen that when distributing sub-interval end at the discontinuity of costate, the numerical precision improves greatly.

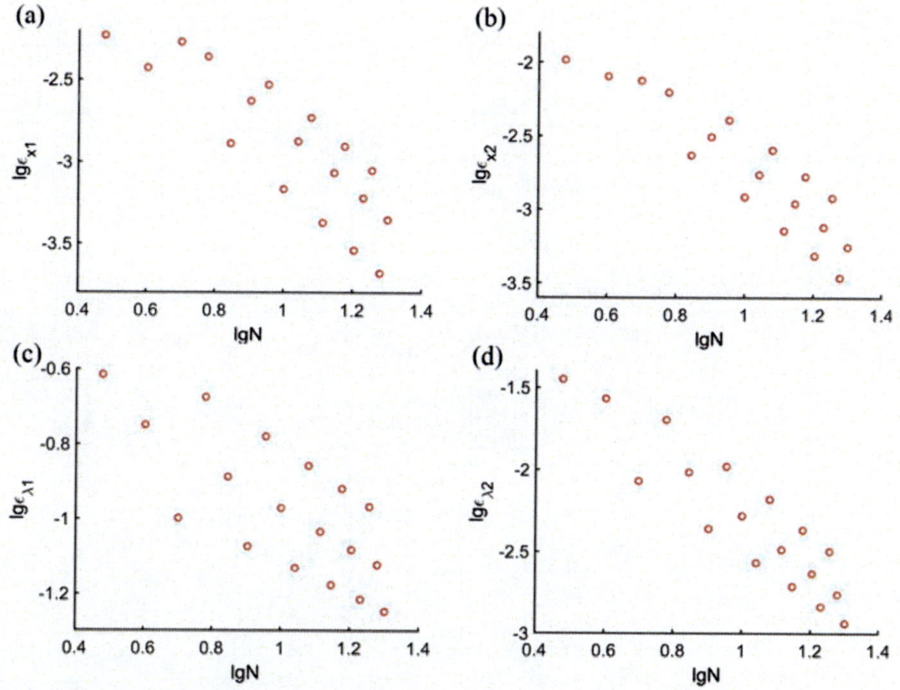


**Fig. 5.9** Numerical precision with fixed $Q$ and varying $N$

**Table 5.1** The precision of costate variables obtained in different numerical methods

| Numerical method | $\varepsilon_{\lambda 1}$ | $\varepsilon_{\lambda 2}$ |
|---|---|---|
| *Adaptive p pseudospectral method* [11] | | |
| LG | × | $O(-1)$ |
| LGR | × | $O(-1)$ |
| *Adaptive h-2 pseudospectral method* [11] | | |
| LG | × | $O(-1)$ |
| LGR | × | $O(-1)$ |
| *Adaptive hp-3 pseudospectral method* [12] | | |
| $\varepsilon = 10^{-2}$ | $1.39 \times 10^{1}$ | $2.02 \times 10^{-1}$ |
| $\varepsilon = 10^{-3}$ | $2.32 \times 10^{1}$ | $2.99 \times 10^{-2}$ |
| $\varepsilon = 10^{-4}$ | $1.30 \times 10^{1}$ | $3.23 \times 10^{-3}$ |
| *SPM* | | |
| $Q = 10,\ N = 10$ | $1.13 \times 10^{-1}$ | $2.90 \times 10^{-3}$ |
| $Q = 15,\ N = 15$ | $6.71 \times 10^{-2}$ | $6.94 \times 10^{-4}$ |
| $Q = 30,\ N = 30$ | $3.98 \times 10^{-2}$ | $9.25 \times 10^{-5}$ |

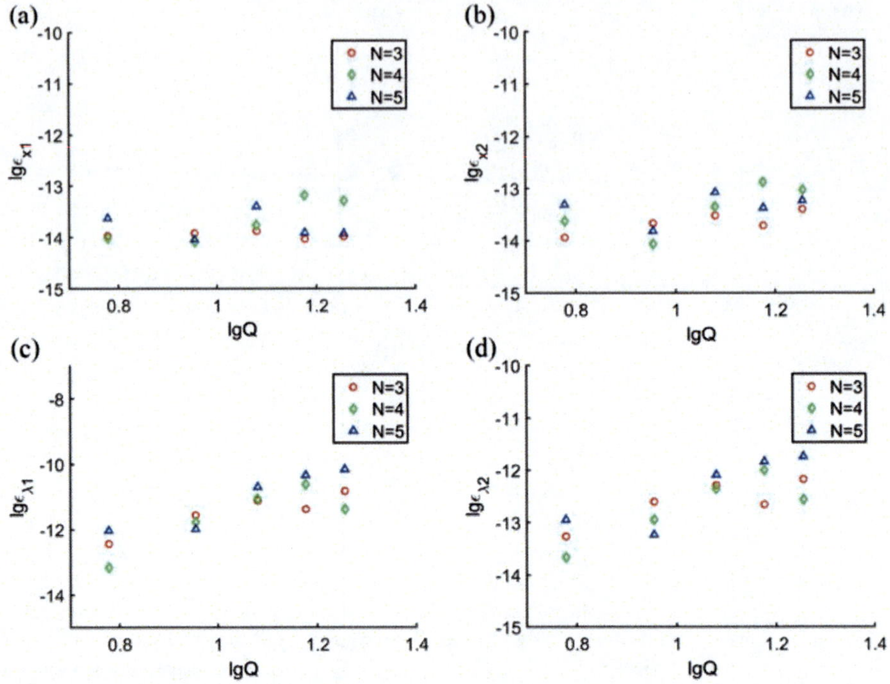


**Fig. 5.10** Numerical precision with fixed $Q$ and varying $N$

## 5.6.2 A Simple Trajectory Planning Problem with Unspecific Terminal Time

In this example, we consider a car-like robot whose motion is described by the following kinematics (it's actually a simplified bicycle model, which will be detailed discussed in Part III in this book)

$$\dot{\mathbf{x}} = \frac{\mathrm{d}}{\mathrm{d}t}\begin{bmatrix} x \\ y \\ \theta \\ v \end{bmatrix} = \begin{bmatrix} v\cos\theta \\ v\cos\theta \\ vu_1/7 \\ u_2 \end{bmatrix} \tag{5.139}$$

where the variable $x$, $y$, and $\theta$ are used to represent the position and orientation of the robot; $v$ represents the velocity; $u_1$ and $u_2$ are the control variables for the system. And the following constraints on state and control variables are imposed

$$|v| - 1 \leq 0 \tag{5.140}$$

$$|u_1| - 2 \le 0 \tag{5.141}$$

$$|u_2| - 1 \le 0 \tag{5.142}$$

In addition, there is an obstacle in the circle shape in the operational environment, and the collision-free condition can be expressed as

$$1 - \left(\frac{x-50}{40}\right)^2 - \left(\frac{y-50}{40}\right)^2 \le 0 \tag{5.143}$$

The initial instant is set as $t_s = 0$, and the robot aims to move from the initial state $\mathbf{x}_s = (0, 0, \pi/2, 0)^{\mathrm{T}}$ to the final state $\mathbf{x}_f = (100, 110, 0, 0)^{\mathrm{T}}$, and the following cost function should be minimized

$$J = 0.1t_f + \int_0^{t_f} (u_1^2 + u_2^2)\mathrm{d}t \tag{5.144}$$

When solving this problem by the algorithm developed in this chapter, the solution parameters are set as follows. First, the time domain is divided into $Q = 5$ sub-intervals and a 30-order LGL approximation is used in each sub-interval; the convergent tolerance in Eq. (5.11), is set as $\varepsilon = 10^{-4}$. Secondly, in the secant method to determine the optimal terminal time, the initial guesses on the terminal time are set as $t_{\mathrm{f}}^{[1]} = 180\mathrm{s}$ and $t_{\mathrm{f}}^{[2]} = 170\mathrm{s}$; the convergent parameters are set as $\varepsilon' = 10^{-3}$ and $\tau = 10^{-3}$. The converged solutions, as illustrated in Figs. 5.11, 5.12, 5.13, and 5.14, are obtained after 4 iterations in the secant method and the optimal terminal time is $t_f = 168.5533$ s.

The series of hollow red circles in Fig. 5.11, represent the outline of the robot, and the blue circle represents the static obstacle. It is seen that the generated path

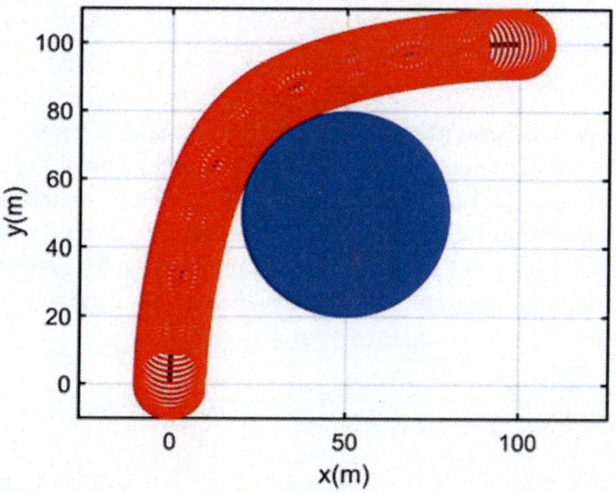


**Fig. 5.11** The optimal trajectory of the robot

Fig. 5.12 Profile of the velocity

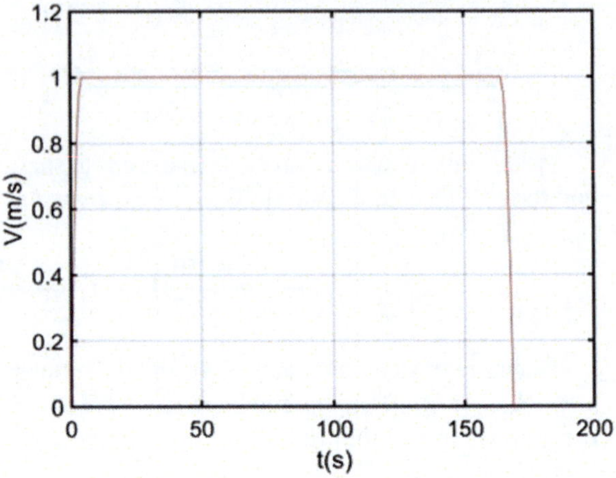


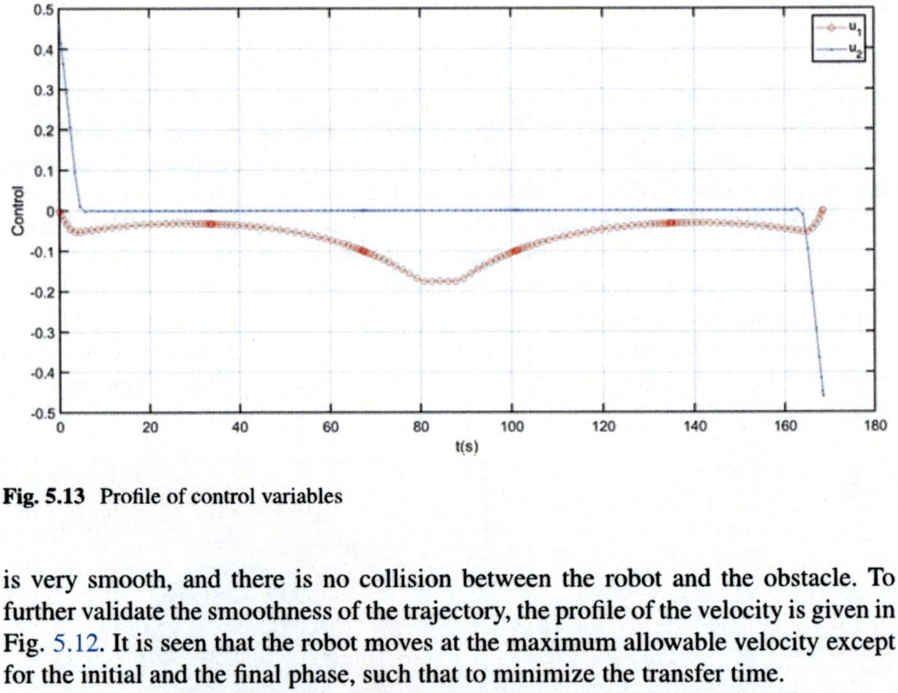


Fig. 5.13 Profile of control variables

is very smooth, and there is no collision between the robot and the obstacle. To further validate the smoothness of the trajectory, the profile of the velocity is given in Fig. 5.12. It is seen that the robot moves at the maximum allowable velocity except for the initial and the final phase, such that to minimize the transfer time.

In Fig. 5.13, the profiles of control variables are presented. It's seen that the control variables are smooth. Finally, the costate variables are given in Fig. 5.14.

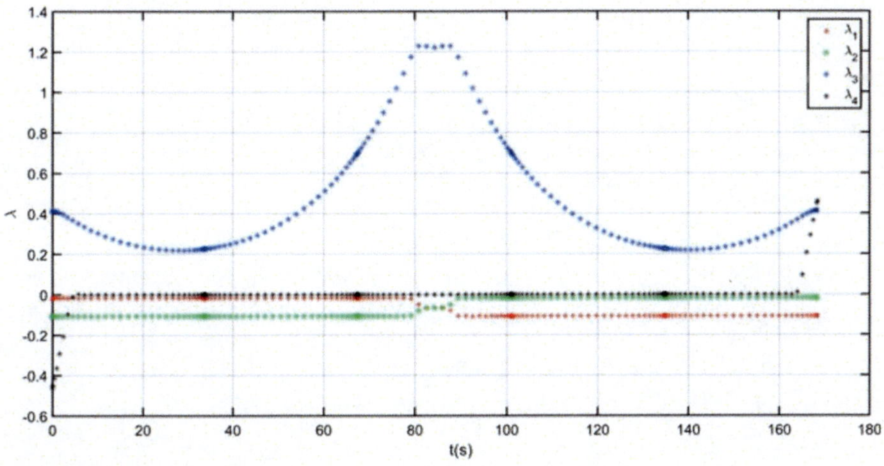


**Fig. 5.14** Profile of costate variables

## References

1. Wang X, Peng H, Zhang S et al (2017) A symplectic pseudospectral method for nonlinear optimal control problems with inequality constraints. ISA Trans 68:335–352
2. Wang X, Peng H, Shi B, Jiang D, Zhang S, Chen B (2019) Optimal vaccination strategy of a constrained time-varying SEIR epidemic model. Commun Nonlinear Sci Numer Simul 67:37–48
3. Luo S, Sun Q, Tao J et al (2016) Trajectory planning and gathering for multiple parafoil systems based on pseudo-spectral method. In: 35th Chinese control conference, 27-29 July 2016. Chengdu, China
4. Li M, Peng H, Zhong W (2016) Optimal control of loose spacecraft formations near libration points with collision avoidance. Nonlinear Dyn 83(4):2241–2261
5. Bryson AE, Ho YC (1975) Applied optimal control. Hemisphere/Wiley, New York
6. Kirk DE (2004) Optimal control theory: an introduction. Dover Publication, New York
7. Li M, Peng H, Zhong W (2015) A symplectic sequence iteration approach for nonlinear optimal control problems with state-control constraints. J Franklin Inst 352(6):2381–2406
8. Li M, Peng H (2016) Solutions of nonlinear constrained optimal control problems using quasilinearization and variational pseudospectral methods. ISA Trans 62:177–192
9. Jiang C, Lin Q, Yu C et al (2012) An exact penalty method for free terminal time optimal control problem with continuous inequality constraints. J Optim Theory Appl 154(1):30–53
10. Korayem MH, Nazemizadeh M, Nohooji HR (2014) Optimal point-to-point motion planning of non-holonomic mobile robots in the presence of multiple obstacles. J Braz Soc Mech Sci Eng 36(1):221–232
11. Darby CL, Garg D, Rao AV (2015) Costate estimation using multiple-interval pseudospectral methods. J Spacecraft Rockets 48(5):856–866
12. Thorvaldsen TP, Huntington GT, Benson DA (2006) Direct trajectory optimization and costate estimation via an orthogonal collocation method. J Guid Control Dyn 29(6):1435–1439

# Chapter 6
# SPM for Time-Delayed Nonlinear Optimal Control Problems

## 6.1 Introduction

Time delays naturally exist in modeling and control for various fields, such as chemical engineering [1, 2], vibration control [3, 4], biological control [5, 6], etc. The existence of time delays makes it hard to obtain analytical solutions for the optimal control problem of time-delay systems, especially for systems involving high nonlinearity and/or complicated constraints. Hence, various numerical methods to solve time-delayed optimal control problems (TDOCPs) are developed.

Studies on TDOCPs initially focus on developing corresponding Pontryagin's maximum principle (PMP) for problems with various features, facilitating the development of indirect methods. Kharatishvili does the pioneering work in this field, where the PMP for TDOCPs with a constant state delay is derived [7]. Later, he derives the PMP for TDOCPs with multiple control delays in [8]. Since then, based on Kharatishvili's work, studies on TDOCPs with more complicated factors, such as time delays in both state and control variables [9], inequality constraints [10], proportional delays [11] are conducted, greatly promoting the development of this field.

A TDOCP generally can be transformed into a Hamiltonian two-point boundary value problem (TPBVP) by the PMP or the variational principle. TPBVPs resulted from TDOCPs feature in the following aspects [8]:

(i) Delayed state and advanced costate are both involved in the TPBVP, i.e., the changing law of the state is influenced by the delayed state while the changing law of costate is influenced by advanced costate;
(ii) Information on state and/or control variables before the initial instant is required to determine the uniqueness of the control problem;
(iii) The TPBVP exhibits a piecewise structure;



X. Wang et al., *Symplectic Pseudospectral Methods for Optimal Control*, Intelligent Systems, Control and Automation: Science and Engineering 97,
https://doi.org/10.1007/978-981-15-3438-6_6

(iv) The TPBVP would be coupled with other algebraic equations when constraints are involved in the problem.

It should be pointed out that once control delays are involved, the optimality conditions would be extremely complicated where many switching structures exhibit. Hence, some researches first eliminate the control variable by variable transformations [3, 4]. However, this trick is only applicable to problems with one control-delay.

From the above discussions, it is seen that the TPBVP resulted from the TDOCP is inherently complicated. Hence, direct methods are much widely used to solve TDOCPs when compared to indirect methods. For example, Elnagar develops a global Chebyshev pseudospectral method to solve nonlinear state delay TDOCPs [12]; Maleki develops a global non-classical pseudospectral method to solve the brachistochrone problem [13]. While it is pointed that the real solutions of a TDOCP have a piecewise structure [14], which suggests that solving TDOCPs under a global interpolation framework may not lead to solutions of high precision. Thus, the local interpolation framework draws much attention in recent years. For example, Perng develops a local LG pseudospectral method to solve linear TDOCPs [15]; Maleki develops a local LGR pseudospectral method to solve nonlinear TDOCPs with time-varying state delays [16], and he recently develops a local LG pseudospectral method to solve nonlinear TDOCPs with inequality constraints in [14]. It is noted that the successive techniques are used in both [14, 16], where the original nonlinear problem is transformed into a series of linear-quadratic problems.

In the rest of this section, we focus on TDOCPs with a single state delay. This is actually the simplest problem in TDOCPs. Inequality constraints can also be considered, but it is not presented in the derivation for simplicity.

## 6.2 Problem Formulation

Consider the following nonlinear optimal control problem with a constant delay in state variables:

**Problem 6.1**
Minimize the following quadratic Bolza type cost functional

$$J = \frac{1}{2}\mathbf{x}^{\mathrm{T}}(t_f)\mathbf{Z}_f\mathbf{x}(t_f) + \frac{1}{2}\int_{t_s}^{t_f} (\mathbf{x}^{\mathrm{T}}\mathbf{P}\mathbf{x} + \mathbf{u}^{\mathrm{T}}\mathbf{R}\mathbf{u})\mathrm{d}t \tag{6.1}$$

subject to the state-delayed system equation

$$\dot{\mathbf{x}} = \mathbf{f}(\mathbf{x}(t), \mathbf{x}(t-\theta), \mathbf{u}(t), t),\ t \in [t_s, t_f] \tag{6.2}$$

and the initial conditions

$$\mathbf{x}(t) \equiv \mathbf{x}_s,\ t \in [t_s - \theta, t_s] \tag{6.3}$$

where $t_s$ and $t_f$ are the fixed initial and terminal time, respectively. $\mathbf{x}(t) \in \mathbb{R}^d$ is the state variable and $\mathbf{x}_s \in \mathbb{R}^d$ represents its initial condition; $\theta \in \mathbb{R}^+$ represents the constant state delay; $\mathbf{u}(t) \in \mathbb{R}^p$ is the control variable; $\mathbf{f} : \mathbb{R}^d \times \mathbb{R}^d \times \mathbb{R}^p \times \mathbb{R} \to \mathbb{R}^d$ is the nonlinear function to describe the system equation; $\mathbf{Z}_f \in \mathbb{R}^{d\times d}$, $\mathbf{P} \in \mathbb{R}^{d\times d}$ and $\mathbf{R} \in \mathbb{R}^{p\times p}$ are weighted matrices on terminal states, processing states and processing control, respectively. Matrices $\mathbf{Z}_f$ and $\mathbf{P}$ are positive semi-definite, while matrix $\mathbf{R}$ is positive definite

Two types of terminal boundary conditions, i.e., the fixed terminal state and the free terminal states are considered. As for the condition of the fixed terminal state, the prescribed terminal state is denoted as $\mathbf{x}_g \in \mathbb{R}^d$.

## 6.3 Problem Transformation

Using the successive convexification (SCvx) techniques, i.e., linearizing the system equation around the nominal solutions, a series of linear-quadratic problems in the following fashion can be obtained:

Minimize

$$J^{[k+1]} = \frac{1}{2}\left(\mathbf{x}^{[k+1]}(t_f)\right)^{\mathrm{T}}\mathbf{Z}_f\mathbf{x}^{[k+1]}(t_f) + \frac{1}{2}\int_{t_s}^{t_f}\left(\left(\mathbf{x}^{[k+1]}\right)^{\mathrm{T}}\mathbf{P}\mathbf{x}^{[k+1]} + \left(\mathbf{u}^{[k+1]}\right)^{\mathrm{T}}\mathbf{R}\mathbf{u}^{[k+1]}\right)\mathrm{d}t \tag{6.4}$$

subject to the linearized state-delayed system equation

$$\begin{aligned}\dot{\mathbf{x}}^{[k+1]}(t) &= \mathbf{A}^{[k]}(t)\mathbf{x}^{[k+1]}(t) + \mathbf{B}^{[k]}(t)\mathbf{u}^{[k+1]}(t)\\ &\quad + \bar{\mathbf{A}}^{[k]}(t)\mathbf{x}^{[k+1]}(t-\theta) + \mathbf{w}^{[k]}(t),\ t \in \left[t_s, t_f\right]\end{aligned} \tag{6.5}$$

and the initial condition

$$\mathbf{x}^{[k+1]}(t) = \mathbf{x}_s,\ t \in [t_s - \theta, t_s] \tag{6.6}$$

where the notation $(\bullet)^k$ represents the result of a variable $(\bullet)$ obtained in the $k$ th iteration. The coefficient matrices involved are express as follows:

$$\mathbf{A}^{[k]}(t) = \left.\frac{\partial \mathbf{f}(\mathbf{x}(t), \mathbf{x}(t-\theta), \mathbf{u}(t), t)}{\partial \mathbf{x}(t)}\right|_{\mathbf{x}^{[k]}(t), \mathbf{x}^{[k]}(t-\theta), \mathbf{u}^{[k]}(t)} \tag{6.7}$$

$$\mathbf{B}^{[k]}(t) = \left.\frac{\partial \mathbf{f}(\mathbf{x}(t), \mathbf{x}(t-\theta), \mathbf{u}(t), t)}{\partial \mathbf{u}(t)}\right|_{\mathbf{x}^{[k]}(t), \mathbf{x}^{[k]}(t-\theta), \mathbf{u}^{[k]}(t)} \tag{6.8}$$

$$\bar{\mathbf{A}}^{[k]}(t) = \left.\frac{\partial \mathbf{f}(\mathbf{x}(t), \mathbf{x}(t-\theta), \mathbf{u}(t), t)}{\partial \mathbf{x}(t-\theta)}\right|_{\mathbf{x}^{[k]}(t), \mathbf{x}^{[k]}(t-\theta), \mathbf{u}^{[k]}(t)} \tag{6.9}$$

$$\begin{aligned}\mathbf{w}^{[k]}(t) = {} & \mathbf{f}\left(\mathbf{x}^{[k]}(t), \mathbf{x}^{[k]}(t-\theta), \mathbf{u}^{[k]}(t), t\right) - \mathbf{A}^{[k]}(t)\mathbf{x}^{[k]}(t) \\ & - \mathbf{B}^{[k]}(t)\mathbf{u}^{[k]}(t) - \bar{\mathbf{A}}^{[k]}(t)\mathbf{x}^{[k]}(t-\theta)\end{aligned} \tag{6.10}$$

In the following iterative solving process, the results obtained in the $k$ th iteration are taken as the reference solutions for the $(k+1)$ th iteration. And the iteration process goes on until the following convergent criterion achieves

$$\left\|\mathbf{c}^{[k+1]} - \mathbf{c}^{[k]}\right\| / \left\|\mathbf{c}^{[k+1]}\right\| \le \varepsilon \tag{6.11}$$

where $\mathbf{c}^{[k]}$ and $\mathbf{c}^{[k+1]}$ are the solutions obtained in the $k$ th and the $(k+1)$ th iterations, respectively. And the variable $\varepsilon$ is a small quantity that represents the convergent tolerance.

To simplify the following derivations, we consider the following problem without the iteration index

**Problem 6.2**
Minimize

$$J = \frac{1}{2}\mathbf{x}^{\mathrm{T}}(t_f)\mathbf{Z}_f\mathbf{x}(t_f) + \frac{1}{2}\int_{t_s}^{t_f} \left(\mathbf{x}^{\mathrm{T}}\mathbf{P}\mathbf{x} + \mathbf{u}^{\mathrm{T}}\mathbf{R}\mathbf{u}\right)\mathrm{d}t \tag{6.12}$$

subject to the linearized state-delayed system equation

$$\dot{\mathbf{x}}(t) = \mathbf{A}(t)\mathbf{x}(t) + \mathbf{B}(t)\mathbf{u}(t) + \bar{\mathbf{A}}(t)\mathbf{x}(t-\theta) + \mathbf{w}(t), \ \text{for } t \in \left[t_s, t_f\right] \tag{6.13}$$

and the initial condition

$$\mathbf{x}(t) = \mathbf{x}_s, \ t \in [t_s - \theta, t_s] \tag{6.14}$$

Then, according to Sect. 3.3, the first-order necessary conditions of Problem 6.2 can be derived. First, by introducing the costate variable $\boldsymbol{\lambda}$, the Hamiltonian of Problem 6.2 can be expressed as

$$\bar{H}(\mathbf{x}, \boldsymbol{\lambda}, \mathbf{u}, t) = \frac{1}{2}\mathbf{x}^{\mathrm{T}}(t)\mathbf{P}\mathbf{x}(t) + \frac{1}{2}\mathbf{u}^{\mathrm{T}}(t)\mathbf{R}\mathbf{u}(t) + \boldsymbol{\lambda}^{\mathrm{T}}(t)\big(\mathbf{A}(t)\mathbf{x}(t) + \mathbf{B}(t)\mathbf{u}(t) + \bar{\mathbf{A}}(t)\mathbf{x}(t-\theta) + \mathbf{w}(t)\big) \tag{6.15}$$

Then according to the optimality condition w.r.t. the control variable, one has

$$\frac{\partial \bar{H}(\mathbf{x}, \boldsymbol{\lambda}, \mathbf{u}, t)}{\partial \mathbf{u}} = \mathbf{0} \tag{6.16}$$

By solving Eq. (6.16), we can express the control variable as a function of state, costate, and parametric variables as follows:

$$\mathbf{u}(t) = -\mathbf{R}^{-1}\mathbf{B}^{\mathrm{T}}(t)\boldsymbol{\lambda}(t) \tag{6.17}$$

Substituting Eq. (6.17), back into Eq. (6.15), the Hamiltonian can be rewritten as follows:

$$H = \frac{1}{2}\mathbf{x}^{\mathrm{T}}(t)\mathbf{P}\mathbf{x}(t) - \frac{1}{2}\boldsymbol{\lambda}^{\mathrm{T}}(t)\mathbf{B}(t)\mathbf{R}^{-1}\mathbf{B}^{\mathrm{T}}(t)\boldsymbol{\lambda}(t) + \boldsymbol{\lambda}^{\mathrm{T}}(t)\big(\mathbf{A}(t)\mathbf{x}(t) + \bar{\mathbf{A}}(t)\mathbf{x}(t-\theta) + \mathbf{w}(t)\big) \tag{6.18}$$

Furthermore, the optimality conditions w.r.t. state and costate variables, which are also called the Hamiltonian canonical equations, can be derived as

$$\dot{\mathbf{x}}(t) = \frac{\partial H}{\partial \boldsymbol{\lambda}(t)} = \mathbf{A}(t)\mathbf{x}(t) + \bar{\mathbf{A}}(t)\mathbf{x}(t-\theta) + \mathbf{w}(t) - \mathbf{B}(t)\mathbf{R}^{-1}\mathbf{B}^{\mathrm{T}}(t)\boldsymbol{\lambda}(t),\ t \in [t_s, t_f] \tag{6.19}$$

$$\begin{aligned}\dot{\boldsymbol{\lambda}}(t) &= -\frac{\partial H}{\partial \mathbf{x}(t)} - \frac{\partial H}{\partial \mathbf{x}(t-\theta)}\chi_{[t_s, t_f-\theta]}(t) \\ &= \begin{cases} -\mathbf{P}\mathbf{x}(t) - \mathbf{A}^{\mathrm{T}}(t)\boldsymbol{\lambda}(t) - \bar{\mathbf{A}}^{\mathrm{T}}(t+\theta)\boldsymbol{\lambda}(t+\theta),\ t \in [t_s, t_f - \theta] \\ -\mathbf{P}\mathbf{x}(t) - \mathbf{A}^{\mathrm{T}}(t)\boldsymbol{\lambda}(t),\ t \in [t_f - \theta, t_f] \end{cases}\end{aligned} \tag{6.20}$$

where the notation $\chi_{[a,b]}(t)$ satisfies

$$\chi_{[a,b]}(t) = \begin{cases} 1, \text{ if } a \le t \le b \\ 0, \text{ otherwise} \end{cases} \tag{6.21}$$

It's noted that the terminal state can be either fixed or free, thus the terminal boundary condition can be derived according to Sect. 3.3. For a fixed terminal state, one has

$$\mathbf{x}(t_s) = \mathbf{x}_s,\ \mathbf{x}(t_f) = \mathbf{x}_g \tag{6.22}$$

and for the free terminal state, one has

$$\mathbf{x}(t_s) = \mathbf{x}_s, \boldsymbol{\lambda}(t_f) = \mathbf{Z}_f \mathbf{x}_f \tag{6.23}$$

Now, the first-order necessary conditions and the boundary conditions for Problem 6.2 are already derived. It's seen that the first-order necessary conditions for this problem are a Hamiltonian two-point boundary value problem where both delayed state variable and advanced costate variable are involved.

**Remark 6.1** Actually, symplectic pseudospectral methods for nonlinear optimal control problems consider non-quadratic cost functional and inequality constraints can be also developed. However, to simplify the derivation, such two factors are not considered herein. For nonlinear problems with these two factors, one can first apply the SCvx technique to transform it into a series of linear-quadratic problems by linearizing the system equation and constraints, meanwhile expand the cost functional up to the second order around the nominal solutions, similar to what is done in Sect. 5.3. The first-order necessary conditions for the linear-quadratic problem is the coupling of a Hamiltonian two-point boundary value problem and a linear complementarity problem. And the symplectic method based on the second kind of generating function can be developed. Interesting readers can further refer to [17, 18]. □

## 6.4 Algorithm Construction

Based on the first-order necessary conditions derived in Sect. 6.3, using the first kind of generating function and the LGL local pseudospectral method, a symplectic pseudospectral method to solve Problem 6.2 is developed. The SPM is mainly divided into 5 steps:

(1) Dividing the time domain into several regular sub-intervals with special requirements on mesh length.
(2) Applying the LGL pseudospectral method to state and costate variables in each sub-interval.
(3) Applying the variational principle to the second kind of generating function to form a system of linear algebraic equations.
(4) Applying the boundary conditions to the system of linear algebraic equations.
(5) Solving the system of linear algebraic equations formed in step (4).

### 6.4.1 *Time Domain Discretization*

Let the variable $T = t_f - t_s$ represent the length of the time domain $\left[t_s, t_f\right]$. Then we divide the time domain $\left[-\theta + t_s, t_f\right]$ into $(Q + q)$ regular sub-intervals. The length

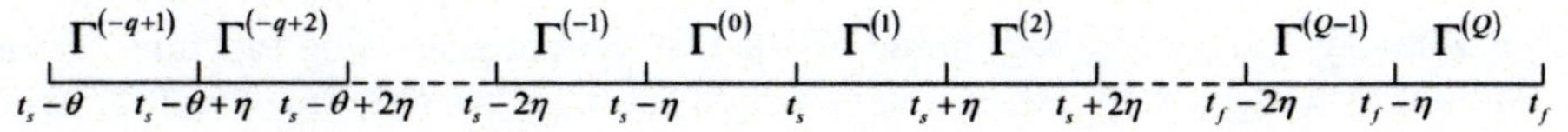


**Fig. 6.1** An illustration of the mesh

of each sub-interval equals to $\eta$ and it satisfies the following conditions

$$\frac{T}{\eta} = Q \in \mathbb{Z}^{+} \quad \text{and} \quad \frac{\theta}{\eta} = q \in \mathbb{Z}^{+} \tag{6.24}$$

Thus, the $j$ th sub-interval is denoted as $\Gamma^{(j)} = \left[t_{j-1}, t_j\right] = [t_s + (j-1)\eta, t_s + j\eta]$, where $j = -q+1, -q+2, \ldots 0, 1, 2, \ldots, Q$. The mesh is illustrated in Fig. 6.1. It is seen that $t_{-q} = t_s - \theta$, $t_0 = t_s$ and $t_Q = t_f$. In addition, sub-intervals $\Gamma^{(-q+1)}$, $\Gamma^{(-q+1)}$, $\Gamma^{(-q+2)}, \ldots, \Gamma^{(-1)}$, $\Gamma^{(0)}$ lie in the range of $[t_s - \theta, t_s]$, while sub-intervals $\Gamma^{(1)}$, $\Gamma^{(2)}, \ldots, \Gamma^{(Q)}$ lie in the range of $\left[t_s, t_f\right]$.

### 6.4.2 Applying the LGL Pseudospectral Method

To apply the LGL pseudospectral method in each sub-interval, the $j$th sub-interval is transformed into a standard time interval $[-1, 1]$ by the following linear transformation

$$\tau^{(j)} = \frac{2}{\eta}\left(t - \frac{t_j + t_{j-1}}{2}\right) \tag{6.25}$$

Thus, the first kind of generating function of the $j$th sub-interval can be written as

$$S^{(j)} = \int_{-1}^{1} \left(\left(\boldsymbol{\lambda}^{(j)}\right)^{\mathrm{T}} \frac{d}{d\tau}\mathbf{x}^{(j)} - \frac{\eta}{2} H\left(\mathbf{x}^{(j)}, \mathbf{x}^{(j-q)}, \boldsymbol{\lambda}^{(j)}, t\right)\right) \mathrm{d}\tau, \; j = 1, 2, \ldots, Q \tag{6.26}$$

Applying the $N$-degree LGL pseudospectral method to the state and costate variables in each sub-interval, then one has

$$\mathbf{x}^{(j)}(\tau) = \sum_{l=0}^{N} \mathbf{x}_l^{(j)} \rho_l^{(j)}(\tau), \; j = -q+1, -q+2, \ldots 0, 1, 2, \ldots, Q \tag{6.27}$$

$$\boldsymbol{\lambda}^{(j)}(\tau) = \sum_{l=0}^{N} \boldsymbol{\lambda}_l^{(j)} \rho_l^{(j)}(\tau), \; j = 1, 2, \ldots, Q \tag{6.28}$$

As in symplectic methods based on the first kind of generating function, for any sub-interval, states at the left end and the right end are taken as independent variables. Hence, for the simplicity of the following derivation, we denote the nonindependent state and costate variables within the $j$th sub-interval by following notations

$$\bar{\mathbf{x}}^{(j)} = \left[\left(\mathbf{x}_1^{(j)}\right)^{\mathrm{T}}, \left(\mathbf{x}_2^{(j)}\right)^{\mathrm{T}}, \ldots, \left(\mathbf{x}_{N-1}^{(j)}\right)^{\mathrm{T}}\right]^{\mathrm{T}} \tag{6.29}$$

$$\bar{\boldsymbol{\lambda}}^{(j)} = \left[\left(\boldsymbol{\lambda}_0^{(j)}\right)^{\mathrm{T}}, \left(\boldsymbol{\lambda}_1^{(j)}\right)^{\mathrm{T}}, \ldots, \left(\boldsymbol{\lambda}_N^{(j)}\right)^{\mathrm{T}}\right]^{\mathrm{T}} \tag{6.30}$$

### 6.4.3 Applying the Variational Principle to the First Kind of Generating Function

In Sect. 6.4.3, unless otherwise stated, the term "$j$ th sub-interval" only refers to $j = 1, 2, \ldots, Q$.

Considering the LGL pseudospectral method applied in Sect. 6.4.2, the first kind of generating function of the $j$th sub-interval can be expressed in the following fashion by the LGL type Gauss integration

$$S^{(j)} = \sum_{k=0}^{N} w_k^{(j)} \left(\left(\boldsymbol{\lambda}_k^{(j)}\right)^{\mathrm{T}} \left(\sum_{l=0}^{N} D_{kl}^{(j)} \mathbf{x}_l^{(j)}\right) - \frac{\eta}{2} H_k^{(j)}\right) \tag{6.31}$$

where $H_k^{(j)}$ represents the value of Hamiltonian at $\tau_k^{(j)}$; $w_k^{(j)}$ is the integral weight as defined in Eq. (3.57); and $\left[D_{kl}^{(j)}\right]$ is the first-order differentiation matrix defined in Eq. (3.68).

Then, according to Eq. (3.35), one has the following stationary conditions

$$\frac{\partial S^{(j)}\left(\mathbf{x}_{j-1}, \bar{\mathbf{x}}^{(j)}, \mathbf{x}_j, \bar{\boldsymbol{\lambda}}^{(j)}\right)}{\partial \bar{\mathbf{x}}^{(j)}} = \mathbf{0} \tag{6.32}$$

$$\frac{\partial S^{(j)}\left(\mathbf{x}_{j-1}, \bar{\mathbf{x}}^{(j)}, \mathbf{x}_j, \bar{\boldsymbol{\lambda}}^{(j)}\right)}{\partial \bar{\boldsymbol{\lambda}}^{(j)}} = \mathbf{0} \tag{6.33}$$

By summing up the first kind of generating function of each sub-interval, the first kind of generating function of the time interval $\left[t_s, t_f\right]$ is obtained as follows:

$$S = \sum_{m=1}^{Q} S^{(m)} \tag{6.34}$$

Applying Eq. (3.35) to Eq. (6.34), one has the following stationary condition

$$\frac{\partial S^{(m)}}{\partial \mathbf{x}_m} + \frac{\partial S^{(m+1)}}{\partial \mathbf{x}_m} = \mathbf{0},\ m = 1, 2, \ldots, Q-1 \tag{6.35}$$

Additionally, according to Eq. (3.36), the states at $t = t_s$ and $t = t_f$ can be obtained as follows:

$$\frac{\partial S(\mathbf{x}_0, \mathbf{x}_Q)}{\partial \mathbf{x}_0} = -\boldsymbol{\lambda}_0 \tag{6.36}$$

$$\frac{\partial S(\mathbf{x}_0, \mathbf{x}_Q)}{\partial \mathbf{x}_Q} = \boldsymbol{\lambda}_Q \tag{6.37}$$

Thus, the relationships in Eqs. (6.32)–(6.33) and (6.35)–(6.37), within the $j$th sub-interval can be put in a unified matrix form as follows:

$$\left(\mathbf{f}_0^x\right)^{(j)} = -\boldsymbol{\lambda}_{j-1} \tag{6.38}$$

$$\left(\mathbf{f}_m^x\right)^{(j)} = \mathbf{0},\ m = 1, 2, \ldots, N-1 \tag{6.39}$$

$$\left(\mathbf{f}_N^x\right)^{(j)} = \boldsymbol{\lambda}_j \tag{6.40}$$

$$\left(\mathbf{f}_m^\lambda\right)^{(j)} = \mathbf{0},\ m = 0, 1, \ldots, N \tag{6.41}$$

where

$$\left(\mathbf{f}_m^x\right)^{(j)} = \frac{\partial S^{(j)}}{\partial \mathbf{x}_m^{(j)}} = \sum_{n=0}^{N} \left(\mathbf{K}_{mn}^{xx}\right)^{(j)} \mathbf{x}_n^{(j)} + \sum_{n=0}^{N} \left(\mathbf{K}_{mn}^{x\lambda}\right)^{(j)} \boldsymbol{\lambda}_n^{(j)} + \sum_{n=0}^{N} \left(\bar{\mathbf{L}}_{mn}^{x\lambda}\right)^{(j)} \boldsymbol{\lambda}_n^{(j+q)} \tag{6.42}$$

$$\left(\mathbf{f}_m^\lambda\right)^{(j)} = \frac{\partial S^{(j)}}{\partial \boldsymbol{\lambda}_m^{(j)}} = \sum_{n=0}^{N} \left(\mathbf{K}_{mn}^{\lambda x}\right)^{(j)} \mathbf{x}_n^{(j)} + \sum_{n=0}^{N} \left(\mathbf{K}_{mn}^{\lambda\lambda}\right)^{(j)} \boldsymbol{\lambda}_n^{(j)} + \sum_{n=0}^{N} \left(\underline{\boldsymbol{L}}_{mn}^{\lambda x}\right)^{(j)} \mathbf{x}_n^{(j-q)} + \left(\boldsymbol{\zeta}_m^\lambda\right)^{(j)} \tag{6.43}$$

From Eqs. (6.42)–(6.43), it is seen that for the first kind of generating function of a certain sub-interval, its partial derivative with respect to state variable within the sub-interval is connected to the costate variable in an advanced sub-interval; while its partial derivative with respect to costate variable within the sub-interval is related to the state variable in a past sub-interval. And the coefficient matrices involved as expressed as follows:

$$(\mathbf{K}_{mn}^{xx})^{(j)} = -\frac{\eta}{2} w_n^{(j)} \mathbf{P}_n^{(j)} \delta_m^n \tag{6.44}$$

$$(\mathbf{K}_{mn}^{x\lambda})^{(j)} = \left((\mathbf{K}_{nm}^{\lambda x})^{(j)}\right)^{\mathrm{T}} = w_n^{(j)} D_{nm}^{(j)} \mathbf{I} - \frac{\eta}{2} w_n^{(j)} (\mathbf{A}_n^{(j)})^{\mathrm{T}} \delta_m^n \tag{6.45}$$

$$(\mathbf{K}_{mn}^{\lambda\lambda})^{(j)} = w_m^{(j)} \frac{\eta}{2} \mathbf{B}_m^{(j)} (\mathbf{R}_m^{(j)})^{-1} (\mathbf{B}_m^{(j)})^{\mathrm{T}} \delta_m^n \tag{6.46}$$

$$(\bar{\mathbf{L}}_{mn}^{x\lambda})^{(j)} = \begin{cases} -\frac{\eta}{2} w_n^{(j)} \left(\bar{\mathbf{A}}_m^{(j+q)}\right)^{\mathrm{T}} \delta_m^n, \ j = 1, 2, \ldots, Q-q \\ \mathbf{0}_{d,d}, \ j = Q-q+1, Q-q+2, \ldots, Q \end{cases} \tag{6.47}$$

$$\left(\underline{\boldsymbol{L}}_{mn}^{\lambda x}\right)^{(j)} = -\frac{\eta}{2} w_m^{(j)} \bar{\mathbf{A}}_m^{(j)} \delta_m^n \tag{6.48}$$

$$(\boldsymbol{\zeta}_m^{\lambda})^{(j)} = -\frac{\eta}{2} w_m^{(j)} \mathbf{w}_m^{(j)} \tag{6.49}$$

To simplify the following derivations, we denote the following notations

$$\bar{\mathbf{L}}^{(j)} = \mathrm{diag}\left((\bar{\mathbf{L}}_{00}^{x\lambda})^{(j)}, (\bar{\mathbf{L}}_{11}^{x\lambda})^{(j)}, \ldots, (\bar{\mathbf{L}}_{NN}^{x\lambda})^{(j)}\right), \ j = 1, 2, \ldots, Q-q \tag{6.50}$$

$$\underline{\boldsymbol{L}}^{(j)} = \mathrm{diag}\left(\left(\underline{\boldsymbol{L}}_{00}^{\lambda x}\right)^{(j)}, \left(\underline{\boldsymbol{L}}_{11}^{\lambda x}\right)^{(j)}, \ldots, \left(\underline{\boldsymbol{L}}_{NN}^{\lambda x}\right)^{(j)}\right), \ j = 1, 2, \ldots, Q \tag{6.51}$$

$$\boldsymbol{\zeta}^{(j)} = \left\{\left((\boldsymbol{\zeta}_0^{\lambda})^{(j)}\right)^{\mathrm{T}}, \left((\boldsymbol{\zeta}_1^{\lambda})^{(j)}\right)^{\mathrm{T}}, \ldots, \left((\boldsymbol{\zeta}_N^{\lambda})^{(j)}\right)^{\mathrm{T}}\right\}^{\mathrm{T}}, \ j = 1, 2, \ldots, Q \tag{6.52}$$

$$\mathbf{X}_0 = \left\{\underbrace{\mathbf{x}_s^{\mathrm{T}}, \mathbf{x}_s^{\mathrm{T}}, \cdots, \mathbf{x}_s^{\mathrm{T}}}_{N+1}\right\}^{\mathrm{T}} \tag{6.53}$$

Thus, the relationships in Eqs. (6.42)–(6.43), can be rewritten into the following compact matrix form

$$\mathbf{K}^{(j)}\begin{bmatrix}\mathbf{x}_{j-1}\\ \bar{\mathbf{x}}^{(j)}\\ \mathbf{x}_j\\ \bar{\boldsymbol{\lambda}}^{(j)}\end{bmatrix}+\underline{\boldsymbol{\Lambda}}^{(j)}\boldsymbol{\beta}^{(j)}+\bar{\boldsymbol{\Lambda}}^{(j)}\begin{bmatrix}\mathbf{0}_{(N+1)d,1}\\ \bar{\boldsymbol{\lambda}}^{(j+q)}\end{bmatrix}\varphi^{(j)}+\begin{bmatrix}\mathbf{0}_{(N+1)d,1}\\ \boldsymbol{\zeta}^{(j)}\end{bmatrix}=\begin{bmatrix}-\boldsymbol{\lambda}_{j-1}\\ \mathbf{0}_{(N-1)d,1}\\ \boldsymbol{\lambda}_j\\ \mathbf{0}_{(N+1)d,1}\end{bmatrix} \tag{6.54}$$

where

$$\boldsymbol{\beta}^{(j)}=\left\{\mathbf{x}_{j-q-1}^{\mathrm{T}}\ \left(\bar{\mathbf{x}}^{(j-q)}\right)^{\mathrm{T}}\ \mathbf{x}_{j-q}^{\mathrm{T}}\ \mathbf{0}_{1,(N+1)d}\right\}^{\mathrm{T}} \tag{6.55}$$

$$\varphi^{(j)}=\begin{cases}1, & j=1,2,\ldots,Q-q\\ 0, & j=Q-q+1,Q-q+2,\ldots,Q\end{cases} \tag{6.56}$$

$$\mathbf{K}^{(j)}=\begin{bmatrix}\mathbf{K}_{11}^{(j)} & \mathbf{K}_{12}^{(j)} & \mathbf{K}_{13}^{(j)} & \mathbf{K}_{14}^{(j)}\\ \mathbf{K}_{21}^{(j)} & \mathbf{K}_{22}^{(j)} & \mathbf{K}_{23}^{(j)} & \mathbf{K}_{24}^{(j)}\\ \mathbf{K}_{31}^{(j)} & \mathbf{K}_{32}^{(j)} & \mathbf{K}_{33}^{(j)} & \mathbf{K}_{34}^{(j)}\\ \mathbf{K}_{41}^{(j)} & \mathbf{K}_{42}^{(j)} & \mathbf{K}_{43}^{(j)} & \mathbf{K}_{44}^{(j)}\end{bmatrix}$$

$$=\left[\begin{array}{c|ccc|c|cccc}
\left(\mathbf{K}_{00}^{xx}\right)^{(j)} & \mathbf{0}_{d,d} & \cdots & \mathbf{0}_{d,d} & \mathbf{0}_{d,d} & \left(\mathbf{K}_{00}^{x\lambda}\right)^{(j)} & \left(\mathbf{K}_{01}^{x\lambda}\right)^{(j)} & \cdots & \left(\mathbf{K}_{0,N-1}^{x\lambda}\right)^{(j)} & \left(\mathbf{K}_{0N}^{x\lambda}\right)^{(j)}\\
\hline
\mathbf{0}_{d,d} & \left(\mathbf{K}_{11}^{xx}\right)^{(j)} & \cdots & \mathbf{0}_{d,d} & \mathbf{0}_{d,d} & \left(\mathbf{K}_{10}^{x\lambda}\right)^{(j)} & \left(\mathbf{K}_{11}^{x\lambda}\right)^{(j)} & \cdots & \left(\mathbf{K}_{1,N-1}^{x\lambda}\right)^{(j)} & \left(\mathbf{K}_{1N}^{x\lambda}\right)^{(j)}\\
\vdots & \vdots & \ddots & \vdots & \vdots & \vdots & \vdots & \ddots & \vdots & \vdots\\
\mathbf{0}_{d,d} & \mathbf{0}_{d,d} & \cdots & \left(\mathbf{K}_{N-1,N-1}^{xx}\right)^{(j)} & \mathbf{0}_{d,d} & \left(\mathbf{K}_{N-1,0}^{x\lambda}\right)^{(j)} & \left(\mathbf{K}_{N-1,1}^{x\lambda}\right)^{(j)} & \cdots & \left(\mathbf{K}_{N-1,N-1}^{x\lambda}\right)^{(j)} & \left(\mathbf{K}_{N-1,N}^{x\lambda}\right)^{(j)}\\
\hline
\mathbf{0}_{d,d} & \mathbf{0}_{d,d} & \cdots & \mathbf{0}_{d,d} & \left(\mathbf{K}_{NN}^{xx}\right)^{(j)} & \left(\mathbf{K}_{N0}^{x\lambda}\right)^{(j)} & \left(\mathbf{K}_{N1}^{x\lambda}\right)^{(j)} & \cdots & \left(\mathbf{K}_{N,N-1}^{x\lambda}\right)^{(j)} & \left(\mathbf{K}_{NN}^{x\lambda}\right)^{(j)}\\
\hline
\left(\mathbf{K}_{00}^{\lambda x}\right)^{(j)} & \left(\mathbf{K}_{01}^{\lambda x}\right)^{(j)} & \cdots & \left(\mathbf{K}_{0,N-1}^{\lambda x}\right)^{(j)} & \left(\mathbf{K}_{0N}^{\lambda x}\right)^{(j)} & \left(\mathbf{K}_{00}^{\lambda\lambda}\right)^{(j)} & \mathbf{0}_{d,d} & \cdots & \mathbf{0}_{d,d} & \mathbf{0}_{d,d}\\
\left(\mathbf{K}_{10}^{\lambda x}\right)^{(j)} & \left(\mathbf{K}_{11}^{\lambda x}\right)^{(j)} & \cdots & \left(\mathbf{K}_{1,N-1}^{\lambda x}\right)^{(j)} & \left(\mathbf{K}_{1N}^{\lambda x}\right)^{(j)} & \mathbf{0}_{d,d} & \left(\mathbf{K}_{11}^{\lambda\lambda}\right)^{(j)} & \cdots & \mathbf{0}_{d,d} & \mathbf{0}_{d,d}\\
\vdots & \vdots & \ddots & \vdots & \vdots & \vdots & \vdots & \ddots & \vdots & \vdots\\
\left(\mathbf{K}_{N-1,0}^{\lambda x}\right)^{(j)} & \left(\mathbf{K}_{N-1,1}^{\lambda x}\right)^{(j)} & \cdots & \left(\mathbf{K}_{N-1,N-1}^{\lambda x}\right)^{(j)} & \left(\mathbf{K}_{N-1,N}^{\lambda x}\right)^{(j)} & \mathbf{0}_{d,d} & \mathbf{0}_{d,d} & \cdots & \left(\mathbf{K}_{N-1,N-1}^{\lambda\lambda}\right)^{(j)} & \mathbf{0}_{d,d}\\
\left(\mathbf{K}_{N0}^{\lambda x}\right)^{(j)} & \left(\mathbf{K}_{N1}^{\lambda x}\right)^{(j)} & \cdots & \left(\mathbf{K}_{N,N-1}^{\lambda x}\right)^{(j)} & \left(\mathbf{K}_{NN}^{\lambda x}\right)^{(j)} & \mathbf{0}_{d,d} & \mathbf{0}_{d,d} & \cdots & \mathbf{0}_{d,d} & \left(\mathbf{K}_{NN}^{\lambda\lambda}\right)^{(j)}
\end{array}\right] \tag{6.57}$$

$$\underline{\boldsymbol{\Lambda}}^{(j)}=\begin{bmatrix}\underline{\boldsymbol{\Lambda}}_{11}^{(j)} & \underline{\boldsymbol{\Lambda}}_{12}^{(j)} & \underline{\boldsymbol{\Lambda}}_{13}^{(j)} & \underline{\boldsymbol{\Lambda}}_{14}^{(j)}\\ \underline{\boldsymbol{\Lambda}}_{21}^{(j)} & \underline{\boldsymbol{\Lambda}}_{22}^{(j)} & \underline{\boldsymbol{\Lambda}}_{23}^{(j)} & \underline{\boldsymbol{\Lambda}}_{24}^{(j)}\\ \underline{\boldsymbol{\Lambda}}_{31}^{(j)} & \underline{\boldsymbol{\Lambda}}_{32}^{(j)} & \underline{\boldsymbol{\Lambda}}_{33}^{(j)} & \underline{\boldsymbol{\Lambda}}_{34}^{(j)}\\ \underline{\boldsymbol{\Lambda}}_{41}^{(j)} & \underline{\boldsymbol{\Lambda}}_{42}^{(j)} & \underline{\boldsymbol{\Lambda}}_{43}^{(j)} & \underline{\boldsymbol{\Lambda}}_{44}^{(j)}\end{bmatrix}=\left[\begin{array}{c|ccc|c|c}
\mathbf{0}_{d,d} & \multicolumn{3}{c|}{\mathbf{0}_{d,(N-1)d}} & \mathbf{0}_{d,d} & \mathbf{0}_{d,(N+1)d}\\
\hline
\mathbf{0}_{(N-1)d,d} & \multicolumn{3}{c|}{\mathbf{0}_{(N-1)d}} & \mathbf{0}_{(N-1)d,d} & \mathbf{0}_{(N-1)\mathrm{d},(N+1)d}\\
\hline
\mathbf{0}_{d,d} & \multicolumn{3}{c|}{\mathbf{0}_{d,(N-1)d}} & \mathbf{0}_{d,d} & \mathbf{0}_{d,(N+1)d}\\
\hline
\left(\underline{\mathbf{L}}_{00}^{\lambda x}\right)^{(j)} & \mathbf{0}_{d,d} & \cdots & \mathbf{0}_{d,d} & \mathbf{0}_{d,d} & \\
\mathbf{0}_{d,d} & \left(\underline{\mathbf{L}}_{11}^{\lambda x}\right)^{(j)} & \cdots & \mathbf{0}_{d,d} & \mathbf{0}_{d,d} & \\
\vdots & \vdots & \ddots & \vdots & \vdots & \mathbf{0}_{(N+1)d,(N+1)d}\\
\mathbf{0}_{d,d} & \mathbf{0}_{d,d} & \cdots & \left(\underline{\mathbf{L}}_{N-1,N-1}^{\lambda x}\right)^{(j)} & \mathbf{0}_{d,d} & \\
\mathbf{0}_{d,d} & \mathbf{0}_{d,d} & \cdots & \mathbf{0}_{d,d} & \left(\underline{\mathbf{L}}_{NN}^{\lambda x}\right)^{(j)} &
\end{array}\right] \tag{6.58}$$

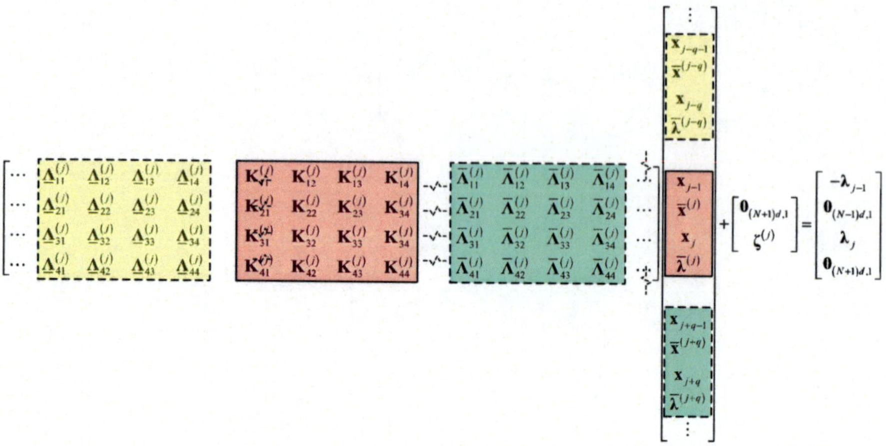

**Fig. 6.2** A graphical illustration of Eq. (6.54)

$$\bar{\boldsymbol{\Lambda}}^{(j)} = \begin{bmatrix} \bar{\boldsymbol{\Lambda}}_{11}^{(j)} & \bar{\boldsymbol{\Lambda}}_{12}^{(j)} & \bar{\boldsymbol{\Lambda}}_{13}^{(j)} & \bar{\boldsymbol{\Lambda}}_{14}^{(j)} \\ \bar{\boldsymbol{\Lambda}}_{21}^{(j)} & \bar{\boldsymbol{\Lambda}}_{22}^{(j)} & \bar{\boldsymbol{\Lambda}}_{23}^{(j)} & \bar{\boldsymbol{\Lambda}}_{24}^{(j)} \\ \bar{\boldsymbol{\Lambda}}_{31}^{(j)} & \bar{\boldsymbol{\Lambda}}_{32}^{(j)} & \bar{\boldsymbol{\Lambda}}_{33}^{(j)} & \bar{\boldsymbol{\Lambda}}_{34}^{(j)} \\ \bar{\boldsymbol{\Lambda}}_{41}^{(j)} & \bar{\boldsymbol{\Lambda}}_{42}^{(j)} & \bar{\boldsymbol{\Lambda}}_{43}^{(j)} & \bar{\boldsymbol{\Lambda}}_{44}^{(j)} \end{bmatrix} = \left[\begin{array}{c|c|c|cccccc} \mathbf{0}_{d,d} & \mathbf{0}_{d,(N-1)d} & \mathbf{0}_{d,d} & \left(\bar{\mathbf{L}}_{00}^{x\lambda}\right)^{(j)} & \mathbf{0}_{d,d} & \cdots & \mathbf{0}_{d,d} & \mathbf{0}_{d,d} \\ \hline & & & \mathbf{0}_{d,d} & \left(\bar{\mathbf{L}}_{11}^{x\lambda}\right)^{(j)} & \cdots & \mathbf{0}_{d,d} & \mathbf{0}_{d,d} \\ \mathbf{0}_{(N-1)d,d} & \mathbf{0}_{(N-1)d} & \mathbf{0}_{(N-1)d,d} & \vdots & \vdots & \ddots & \vdots & \vdots \\ & & & \mathbf{0}_{d,d} & \mathbf{0}_{d,d} & \cdots & \left(\bar{\mathbf{L}}_{N-1,N-1}^{x\lambda}\right)^{(j)} & \mathbf{0}_{d,d} \\ \hline \mathbf{0}_{d,d} & \mathbf{0}_{d,(N-1)d} & \mathbf{0}_{d,d} & \mathbf{0}_{d,d} & \mathbf{0}_{d,d} & \cdots & \mathbf{0}_{d,d} & \left(\bar{\mathbf{L}}_{NN}^{x\lambda}\right)^{(j)} \\ \hline \mathbf{0}_{(N+1)d,d} & \mathbf{0}_{(N+1)d,(N-1)d} & \mathbf{0}_{(N+1)d,d} & \multicolumn{5}{c}{\mathbf{0}_{(N+1)d,(N+1)d}} \end{array}\right] \tag{6.59}$$

To clearly demonstrate the structure of Eq. (6.54), an illustration is given in Fig. 6.2. From Fig. 6.2, we can draw the following three conclusions:

(1) The yellow part in Fig. 6.2, shows the influence of variables in $\Gamma^{(j-q)}$ on $\Gamma^{(j)}$. If $\Gamma^{(j-q)}$ lies in the range of $[t_s - \theta, t_s]$, the state variables therein equal to the prescribed initial conditions, i.e., $\mathbf{x}_l^{(j-q)} = \mathbf{x}_s$, with $l = 1, 2, \ldots, N$ and $j = 1, 2, \ldots, q$; and the costate variables therein are meaningless. Hence, for the $j$ th sub-interval, if $j \in \{1, 2, \ldots, q\}$, then the variable $\boldsymbol{\beta}^{(j)}$ in Eq. (6.54), becomes a constant vector as follows:

$$\boldsymbol{\beta}^{(j)} = \left\{ \mathbf{X}_0^{\mathrm{T}} \; \mathbf{0}_{1,(N+1)d} \right\}^{\mathrm{T}} \tag{6.60}$$

(2) The red part in Fig. 6.2, shows the influence of variables in $\Gamma^{(j)}$ on itself. And neither delayed nor advanced variables are involved.

(3) The green part in Fig. 6.2, shows the influence of variables in $\Gamma^{(j+q)}$ on $\Gamma^{(j)}$. For any $j > Q - q$, $\Gamma^{(j+q)}$ lies on the right-hand side of $t_f$, and both the state and costate variables therein are meaningless. Hence, for the $j$th sub-interval, if $j \in \{Q - q + 1, Q - q + 2, \ldots, Q\}$, then the variable $\varphi^{(j)}$ in Eq. (6.54), equals zeros.

By taking the state and costate variables at all LGL nodes to construct the following unknown variable list

$$\boldsymbol{\chi} = \left\{\mathbf{x}_0^{\mathrm{T}}, (\bar{\mathbf{x}}^{(1)})^{\mathrm{T}}, \left(\bar{\boldsymbol{\lambda}}^{(1)}\right)^{\mathrm{T}}; \mathbf{x}_1^{\mathrm{T}}, (\bar{\mathbf{x}}^{(2)})^{\mathrm{T}}, \left(\bar{\boldsymbol{\lambda}}^{(2)}\right)^{\mathrm{T}}; \ldots; \mathbf{x}_{Q-1}^{\mathrm{T}}, (\bar{\mathbf{x}}^{(Q)})^{\mathrm{T}}, \left(\bar{\boldsymbol{\lambda}}^{(Q)}\right)^{\mathrm{T}}; \mathbf{x}_Q^{\mathrm{T}}\right\}^{\mathrm{T}} \tag{6.61}$$

thus, the relationships in Eqs. (6.32)–(6.33) and Eqs. (6.35)–(6.37), can be written into the following matrix form

$$\mathbf{K}\boldsymbol{\chi} + \boldsymbol{\Lambda}\boldsymbol{\chi} + \boldsymbol{\xi} + \boldsymbol{\zeta} = \boldsymbol{\gamma}, \text{ or } \boldsymbol{\Omega}\boldsymbol{\chi} = \mathbf{r} \tag{6.62}$$

where

$$\boldsymbol{\Omega} = \mathbf{K} + \boldsymbol{\Lambda} \tag{6.63}$$

$$\mathbf{r} = \boldsymbol{\gamma} - \boldsymbol{\xi} - \boldsymbol{\zeta} \tag{6.64}$$

$$\boldsymbol{\gamma} = \{-\boldsymbol{\lambda}_0^{\mathrm{T}}, \mathbf{0}_{1,((2N+1)Q-1)d}, \boldsymbol{\lambda}_Q^{\mathrm{T}}\}^{\mathrm{T}} \tag{6.65}$$

$$\boldsymbol{\xi} = \left\{\mathbf{0}_{1,Nd}, \left(\underline{\boldsymbol{L}}^{(1)}\mathbf{x}_0\right)^{\mathrm{T}}, \mathbf{0}_{1,Nd}, \left(\underline{\boldsymbol{L}}^{(2)}\mathbf{x}_0\right)^{\mathrm{T}}, \ldots, \mathbf{0}_{1,Nd}, \left(\underline{\boldsymbol{L}}^{(q)}\mathbf{x}_0\right)^{\mathrm{T}}, \mathbf{0}_{1,((2N+1)(Q-q)+1)d}\right\}^{\mathrm{T}} \tag{6.66}$$

$$\boldsymbol{\zeta} = \left\{\mathbf{0}_{1,Nd}, (\boldsymbol{\zeta}^{(1)})^{\mathrm{T}}, \mathbf{0}_{1,Nd}, (\boldsymbol{\zeta}^{(2)})^{\mathrm{T}}, \ldots, \mathbf{0}_{1,Nd}, (\boldsymbol{\zeta}^{(Q)})^{\mathrm{T}}, \mathbf{0}_{1,d}\right\}^{\mathrm{T}} \tag{6.67}$$

$$\mathbf{K} = \begin{bmatrix} \ddots & & & & & & & & & & \\ & \overbrace{\mathbf{K}_{33}^{(j-1)} + \mathbf{K}_{11}^{(j)}}^{(2N+2)d \text{ columns}} & \mathbf{K}_{12}^{(j)} & \mathbf{K}_{14}^{(j)} & \mathbf{K}_{13}^{(j)} & & & & & & \\ & \mathbf{K}_{21}^{(j)} & \mathbf{K}_{22}^{(j)} & \mathbf{K}_{24}^{(j)} & \mathbf{K}_{23}^{(j)} & & & & & & \\ & \mathbf{K}_{41}^{(j)} & \mathbf{K}_{42}^{(j)} & \mathbf{K}_{44}^{(j)} & \mathbf{K}_{43}^{(j)} & & & & & & \\ & \mathbf{K}_{31}^{(j)} & \mathbf{K}_{32}^{(j)} & \mathbf{K}_{34}^{(j)} & \mathbf{K}_{33}^{(j)} + \mathbf{K}_{11}^{(j+1)} & \mathbf{K}_{12}^{(j+1)} & \mathbf{K}_{14}^{(j+1)} & \mathbf{K}_{13}^{(j+1)} & & & \\ & & & & \mathbf{K}_{21}^{(j+1)} & \mathbf{K}_{22}^{(j+1)} & \mathbf{K}_{24}^{(j+1)} & \mathbf{K}_{23}^{(j+1)} & & & \\ & & & & \mathbf{K}_{41}^{(j+1)} & \mathbf{K}_{42}^{(j+1)} & \mathbf{K}_{44}^{(j+1)} & \mathbf{K}_{43}^{(j+1)} & & & \\ & & & & \mathbf{K}_{31}^{(j+1)} & \mathbf{K}_{32}^{(j+1)} & \mathbf{K}_{34}^{(j+1)} & \mathbf{K}_{33}^{(j+1)} + \mathbf{K}_{11}^{(j+2)} & \mathbf{K}_{12}^{(j+2)} & \mathbf{K}_{14}^{(j+2)} & \mathbf{K}_{13}^{(j+2)} \\ & & & & & & & \mathbf{K}_{21}^{(j+2)} & \mathbf{K}_{22}^{(j+2)} & \mathbf{K}_{24}^{(j+2)} & \mathbf{K}_{23}^{(j+2)} \\ & & & & & & & \mathbf{K}_{41}^{(j+2)} & \mathbf{K}_{42}^{(j+2)} & \mathbf{K}_{44}^{(j+2)} & \mathbf{K}_{43}^{(j+2)} \\ & & & & & & & \mathbf{K}_{31}^{(j+2)} & \mathbf{K}_{32}^{(j+2)} & \mathbf{K}_{34}^{(j+2)} & \mathbf{K}_{33}^{(j+2)} + \mathbf{K}_{11}^{(j+3)} \\ & & & & & & & & & & \ddots \end{bmatrix} \tag{6.68}$$

(First diagonal block: $(2N+2)d$ rows × $(2N+2)d$ columns.)

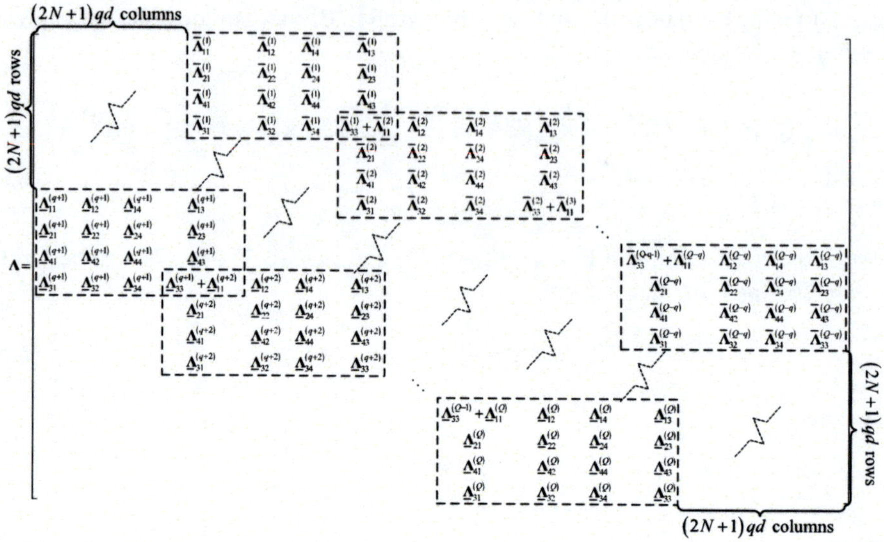


(6.69)

**Remark 6.2** From the expression of matrices $\mathbf{K}$ and $\mathbf{\Lambda}$, we find them to be sparse and symmetric. Hence, we have that matrix $\mathbf{\Omega}$ is sparse and symmetric. And the half bandwidth of matrix $\mathbf{\Omega}$ is $b = ((2N+1)q + N + 1)d$. □

### *6.4.4 Imposing the Boundary Conditions*

In this section, we discuss how to apply the boundary conditions. The system of linear algebraic equations constructed in Eq. (6.62) should be modified according to the boundary conditions provided in Eqs. (6.22) or (6.23).

(1) for the fixed terminal state

When the problem is with the fixed terminal state, Eq. (6.62) is modified as

$$\mathbf{\Omega}_g \boldsymbol{\chi} = \mathbf{r}_g \tag{6.70}$$

where

$$\mathbf{\Omega}_g = \begin{bmatrix} [\mathbf{I}_d \ \mathbf{0}_{d,(2N+1)Qd}] \\ \mathbf{\Omega}[d+1:(2N+1)Qd] \\ [\mathbf{0}_{d,(2N+1)Qd} \ \mathbf{I}_d] \end{bmatrix} \tag{6.71}$$

$$\mathbf{r}_g = \left\{\mathbf{x}_s^{\mathrm{T}}, (\mathbf{r}[d+1:(2N+1)Qd])^{\mathrm{T}}, \mathbf{x}_f^{\mathrm{T}}\right\}^{\mathrm{T}} \tag{6.72}$$

(2) for the free terminal state

As for this condition, Eq. (6.62) is modified as follows:

$$\boldsymbol{\Omega}_f \boldsymbol{\chi} = \mathbf{r}_f \tag{6.73}$$

$$\boldsymbol{\Omega}_f = \begin{bmatrix} \left[\mathbf{I}_d \ \mathbf{0}_{d,(2N+1)Qd}\right] \\ \boldsymbol{\Omega}[d+1:(2N+1)Qd] \\ \boldsymbol{\Omega}_f^{\mathrm{ter}} = \boldsymbol{\Omega}[(2N+1)Qd+1:((2N+1)Q+1)d] - \left[\mathbf{0}_{d,(2N+1)Qd} \ \mathbf{Z}_f\right] \end{bmatrix} \tag{6.74}$$

$$\mathbf{r}_f = \left\{\mathbf{x}_s^{\mathrm{T}}, (\mathbf{r}[d+1:(2N+1)Qd])^{\mathrm{T}}, \mathbf{0}_{1,d}\right\}^{\mathrm{T}} \tag{6.75}$$

For the simplicity of the following derivations, Eqs. (6.70) and (6.73) are written into a uniform fashion as follows:

$$\boldsymbol{\Omega}_\Delta \boldsymbol{\chi} = \mathbf{r}_\Delta \tag{6.76}$$

where the subscript $\Delta \in \{f, g\}$ denotes the type of terminal boundary condition.

### *6.4.5 Solving the System of Linear Algebraic Equations*

By solving the system of linear algebraic equations Eq. (6.76), state and costate at all LGL nodes are obtained. Then according to Eq. (6.17), the control variable is solved. Thus, Problem 6.2 is finally solved.

## 6.5 Numerical Example

The SPM developed in this section is applied to solve some examples. Consider the convergent criterion Eq. (6.11), we set $\varepsilon = 10^{-10}$, and state variables at all LGL nodes are taken as **c**.

### 6.5.1 The Optimal Control Problem of a Spring-Mass System with a Delayed Damper

Consider the following problem

Minimize

$$J = 5x_1^2(2) + \frac{1}{2}\int_0^2 u^2 \mathrm{d}t \tag{6.77}$$

subject to the linear system equations

$$\begin{cases} \dot{x}_1(t) = x_2(t) \\ \dot{x}_2(t) = -x_1(t) - x_2(t-1) + u(t) \end{cases} \tag{6.78}$$

and the boundary conditions

$$\begin{cases} x_1(t) = 10 \\ x_2(t) = 0 \end{cases}, \text{ for } \quad -1 \le t \le 0 \tag{6.79}$$

Due to the existence of analytical solutions, this problem is often taken as the benchmark for numerical methods to solve state-delayed optimal control problems. And the optimal control has the following analytical solution

$$u^*(t) = \begin{cases} \lambda \sin(2-t) + \frac{\lambda}{2}(1-t)\sin(t-1), & \text{for} \quad 0 \le t \le 1 \\ \lambda \sin(2-t), \quad \text{for} \quad 1 < t \le 2 \end{cases} \tag{6.80}$$

where the parameter $\lambda \approx 2.559938203$. We denote the relative error of control variable as

$$\varepsilon_u = \|u - u^*\|_2 / \|u^*\|_2 \tag{6.81}$$

When setting parameters in the SPM as $Q = 10$ and $N = 5$, convergent solutions can be obtained within 0.005 s, and the obtained cost functional is $J = 3.399118238$. The convergent solutions are plotted in Fig. 6.3, 6.4, and 6.5. From Fig. 6.5, it is seen that the numerical solution of control variable is in good agreement with the analytical one, and the relative error of control variable is $\varepsilon_u = 3.3890e^{-6}$.

To test the influence of the number of sub-intervals and the approximation degree of LGL pseudospectral method used in each sub-interval, we plotted the relative error of control variable under two conditions. In Case I, we fix $N$ while tuning $Q$, where $N \in \{k\}_{k=2}^{7}$ and $Q \in \{2k\}_{k=1}^{7}$; In Case II, we fix $Q$ while tuning $N$, where $Q \in \{2k\}_{k=1}^{6}$ and $N \in \{k\}_{k=2}^{7}$. It is seen that the numerical precision improves as $N$

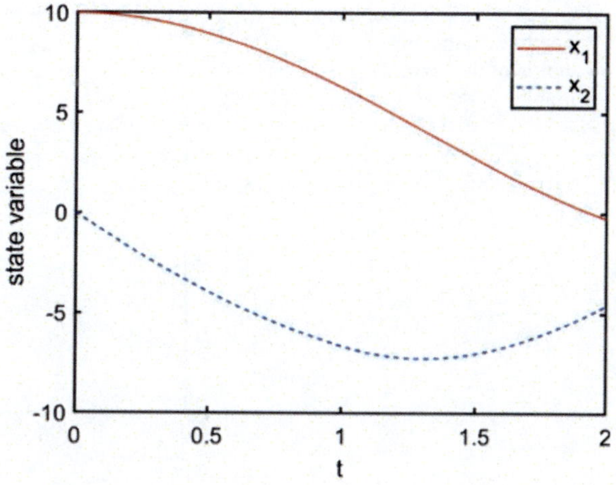


**Fig. 6.3** Numerical solutions of state variables

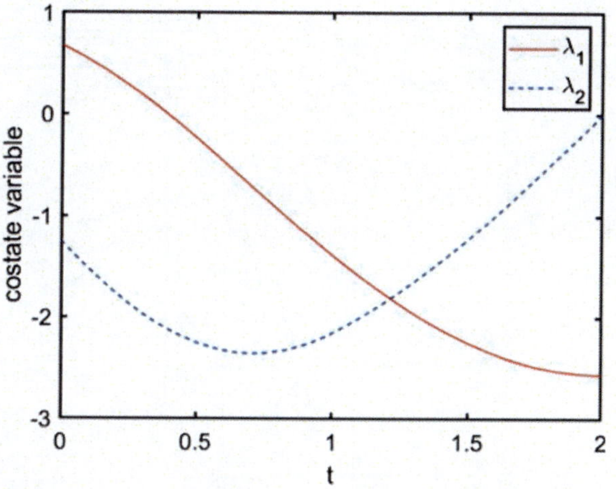


**Fig. 6.4** Numerical solutions of costate variables

and/or $Q$ increases. On the one hand, as illustrated in Fig. 6.6, the SPM exhibits the property of linear convergence when fixing $N$ while tuning $Q$; on the other hand, as illustrated in Fig. 6.7, the SPM shows the property of exponential convergence when fixing $Q$ while tuning $N$, which is similar to traditional pseudospectral methods for unconstrained problems.

This problem is also studied in [14, 19–21]. The solutions of control at different time nodes obtained by the averaging approximation method, the ShLG method and the SPM, together with the analytical solutions, are reported in Table 6.1. The ShLG method in [14], is actually a direct pseudospectral method. And the algorithm parameters therein, i.e., $K$ and $N$ have similar meanings to $Q$ and $N$ in the SPM. From Table 6.1, it is seen that the more accurate numerical solutions are obtained as $N$ increases. In addition, it is noted that when dividing the time domain into

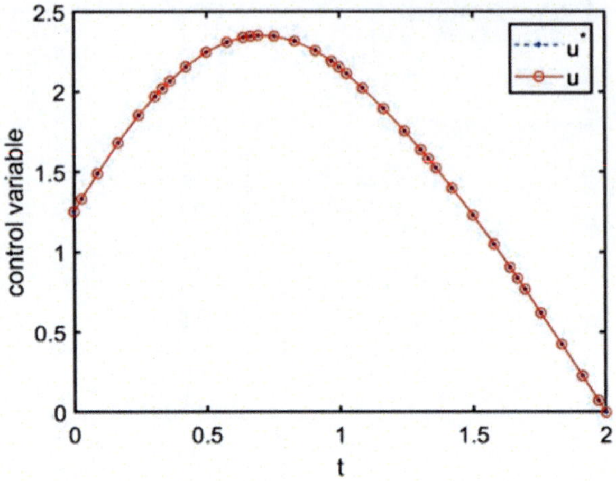


**Fig. 6.5** Numerical solution together with the analytical solution of the control variable

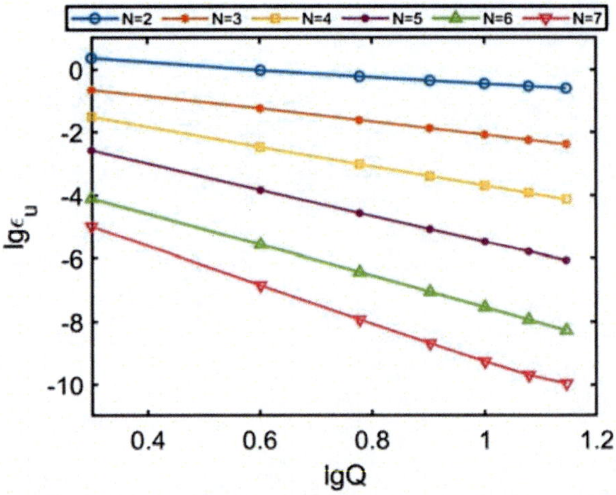


**Fig. 6.6** Numerical precision with fixed $N$ and varying $Q$

two sub-intervals in both the SPM and the ShLG method, solutions obtained by the SPM with $N = 10$ already have the same precision with that obtained by the ShLG method with $N = 14$. This demonstrates that the SPM calls for less computational cost compared to the ShLG method when the same numerical precision is required.

The computational result of cost functional obtained by different numerical methods, together with the analytical solution, are listed in Table 6.2. It is seen that the SPM has better precision when compared with other numerical methods. And one may find that increasing $N$ could more efficiently improve the numerical precision when compared with increasing $Q$.

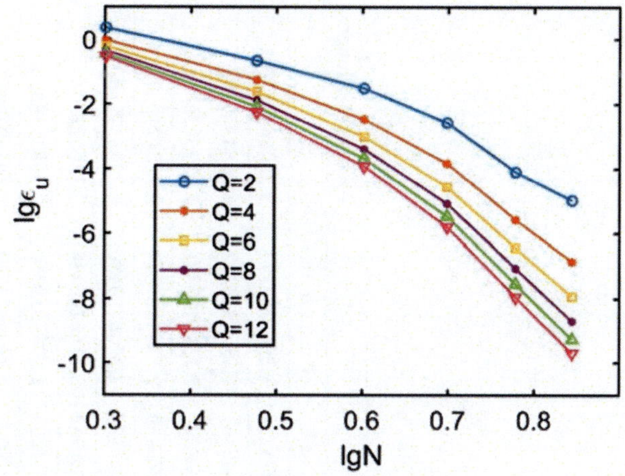


**Fig. 6.7** Numerical precision with fixed $Q$ and varying $N$

### 6.5.2 *Trajectory Planning of a Vehicle with the Steering Delay*

Consider the planar trajectory planning problem of an airplane with steering delay [22]. It is assumed that the airplane moves at a constant velocity $v_0$. Its position is denoted as $(x, y)$, and the orientation is denoted as $\theta$. When applied the control input $u$ to the steering angle $\delta$, orientation changing exhibits with a delay due to the effect such as mechanical friction and aerodynamics. Such a delay effect is modeled by a delay effect in the state variable. When selecting the state space as $\mathbf{x} = [x, y, \theta, \delta]^{\mathrm{T}}$, the system equation can be written as

$$\begin{cases} \dot{x}(t) = v_0 \cos\theta(t) \\ \dot{y}(t) = v_0 \sin\theta(t) \\ \dot{\theta}(t) = c_0 v_0 \delta(t - \tau) \\ \dot{\delta}(t) = u(t) \end{cases} \tag{6.82}$$

where $c_0 > 0$ is a constant parameter and is set as $c_0 = 1$ in this example; $\tau \geq 0$ is the parameter to model the delay effect, and system (6.82), degrades to a non-delay system when $\tau = 0$. In this example, we set the constant velocity as $v_0$=100 m/s, the initial state is $\mathbf{x}(t_s) = \left[0, 0, \pi/4, 5e - 4\right]^{\mathrm{T}}$, and the desired final state is $\mathbf{x}\left(t_f\right) = \left[1500, 1000, \pi/20, 0\right]^{\mathrm{T}}$ with the initial and final time as $t_s = 0$ and $t_f = 19$, respectively. The optimal control problem aims at minimizing the following cost functional

$$J = \int_{t_s}^{t_f} u^2 \mathrm{d}t \tag{6.83}$$

**Table 6.1** Numerical solutions of control at different time nodes

| Time node | Averaging approximation method [21] | ShLG method [14] | SPM | | | Analytical solutions |
|---|---|---|---|---|---|---|
| | | $(K = 2,\ N = 14)$ | $Q = 2,\ N = 5$ | $Q = 2,\ N = 8$ | $Q = 2,\ N = 10$ | |
| 0.00 | 1.2259 | 1.2506884177 | 1.2479394681 | 1.2506875218 | 1.2506884177 | 1.2506884177 |
| 0.20 | 1.7261 | 1.7584349298 | 1.7598741012 | 1.7584346461 | 1.7584349298 | 1.7584349298 |
| 0.41 | 2.0828 | 2.1393125851 | 2.1386810297 | 2.1393128248 | 2.1393125851 | 2.1393125851 |
| 0.61 | 2.2694 | 2.3284279274 | 2.3274216874 | 2.3284276859 | 2.3284279274 | 2.3284279274 |
| 0.81 | 2.2684 | 2.3306379180 | 2.3325149076 | 2.3306381677 | 2.3306379180 | 2.3306379180 |
| 1.02 | 2.0889 | 2.1260220436 | 2.1200172442 | 2.1260221362 | 2.1260220436 | 2.1260220436 |
| 1.22 | 1.7746 | 1.8003519350 | 1.8039489012 | 1.8003521202 | 1.8003519350 | 1.8003519350 |
| 1.42 | 1.3785 | 1.4029074762 | 1.4002240686 | 1.4029073042 | 1.4029074762 | 1.4029074762 |
| 1.62 | 0.9443 | 0.9495335234 | 0.9477236423 | 0.9495337072 | 0.9495335234 | 0.9495335234 |
| 1.82 | 0.4536 | 0.4583046654 | 0.4625444500 | 0.4583044705 | 0.4583046654 | 0.4583046654 |
| 2.00 | 0.0 | 0.0000000000 | −0.0100058783 | −0.0000006713 | 0.0000000000 | 0.0000000000 |

**Table 6.2** Cost functional obtained by different numerical methods

| Numerical methods | Cost functional $J$ |
|---|---|
| Averaging approximation method [21] | **3**.2587 |
| Linear Legendre multiwavelets method [20] | **3**.43254 |
| Hybrid block-pulse and Bernoulli polynomials method [19] | **3**.21663 |
| *The SPM* | |
| $Q = 2, \ N = 6$ | **3.399**143798 |
| $Q = 6, \ N = 6$ | **3.399118**237 |
| $Q = 2, \ N = 10$ | **3.399118063** |
| Analytical solutions | **3.399118063** |

If the state delay is omitted in the trajectory design process, i.e., the condition where $\tau = 0$ can be solved by the SPM developed in Sect. 6.5, and the obtained control variable is denoted as $u^{\otimes}(t)$. When applying $u^{\otimes}(t)$ to the airplane with delay effects, the actual trajectory of the airplane is plotted in Fig. 6.8. In Fig. 6.8, the red line represents the desired trajectory under the ideal condition. It is seen that the actual trajectory deviates from the desired trajectory even when the state delay is one second, resulting in the final position deviation of about 100 m. And the final position deviation increases as the state delay goes larger. For autonomous vehicles in a complex operational environment, if the actual trajectory much deviates from the designed ideal trajectory, it may result in not only mission failing, but also the

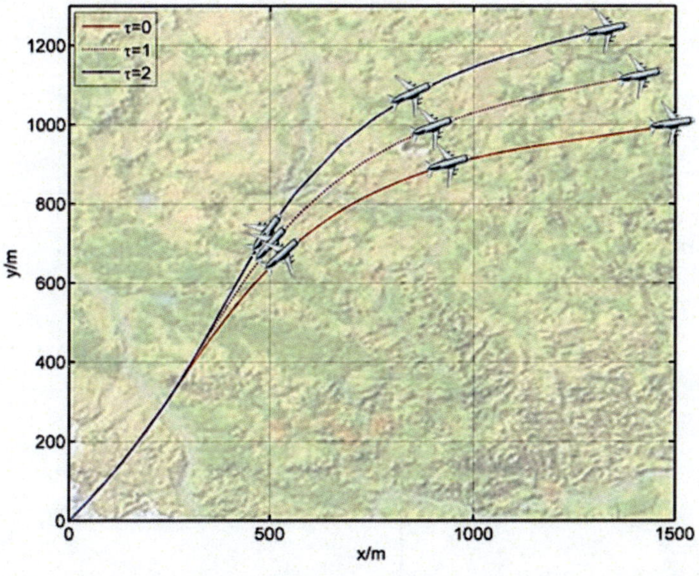


**Fig. 6.8** The real trajectory of the airplane with different time delays when $u^{\otimes}(t)$ is applied

unpredictable damage to the vehicle itself. This enlightens us that it is necessary to consider the inherent delay effect of the vehicle when conducting trajectory planning.

Then, we use the SPM to solve the problem considering the state-delay effect. In the SPM, one sets $Q = 10$ and $N = 10$, and the convergent tolerance in the SCvx procedure is set as $\varepsilon = 10^{-10}$. The computational results for state and control variables with different state delays are plotted in Figs. 6.9, 6.10, 6.11, and 6.12. It is seen that the control variable keeps constant during the time interval $\left[t_f - \tau, t_f\right]$ since state delay exists. And such the constant much closely approaches to zero as the state delay increases. For each state variable, it shows similar variation trending under different state delays.

This example is also solved by the Homotopy Shooting Method (HSM) in [22]. The cost functionals obtained by the HSM and the SPM under different state delays are reported in Table 6.3. From Table 6.3, it is seen that the results obtained by the SPM are slightly smaller.

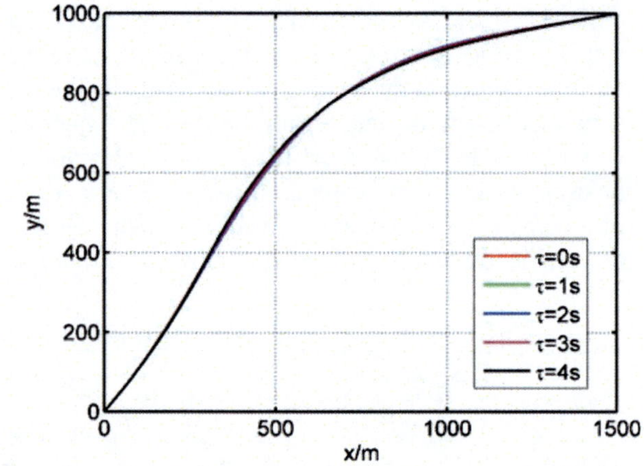


**Fig. 6.9** Optimal trajectories under different state delays

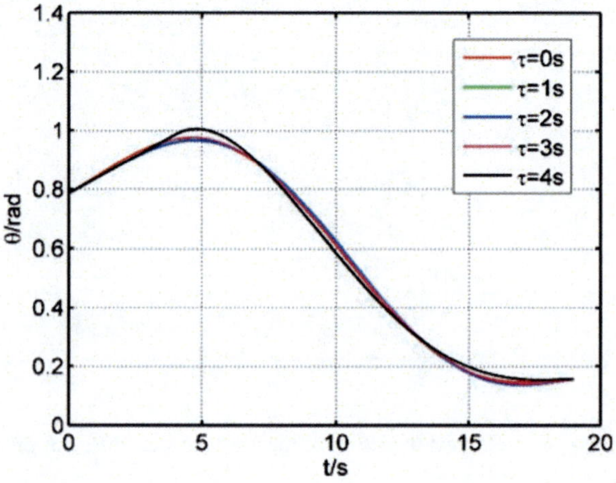


**Fig. 6.10** Optimal orientations under different state delays

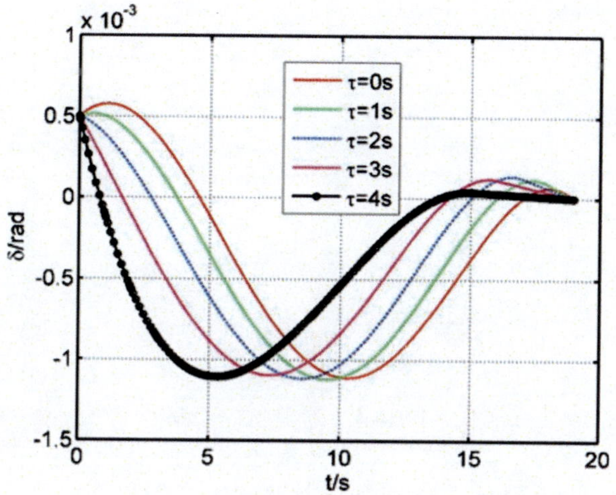


**Fig. 6.11** Optimal steering angles under different state delays

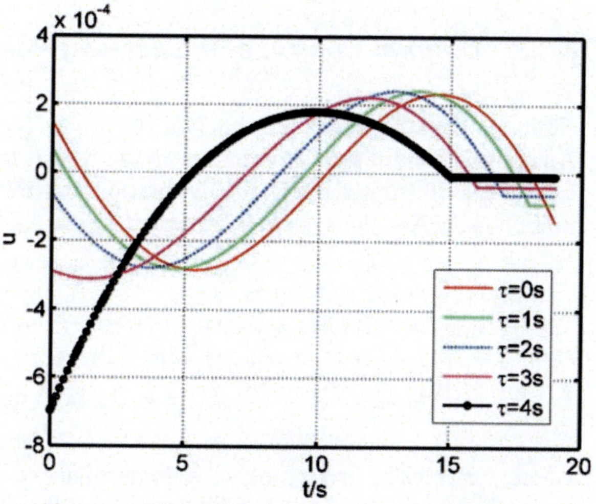


**Fig. 6.12** Optimal control inputs under different state delays

**Table 6.3** Cost functionals obtained by different numerical methods

| State-delay τ/s | Cost functional | | $(J_2-J_1)/J_2 \times$ 100% (%) |
|---|---|---|---|
| | SPM, $J_1$ | HSM [22], $J_2$ | |
| 0 | 6.09062e-7 | 6.09179e-7 | 0.01921 |
| 2 | 6.04701e-7 | 6.05183e-7 | 0.07965 |
| 4 | 8.68535e-7 | 8.68821e-7 | 0.03292 |

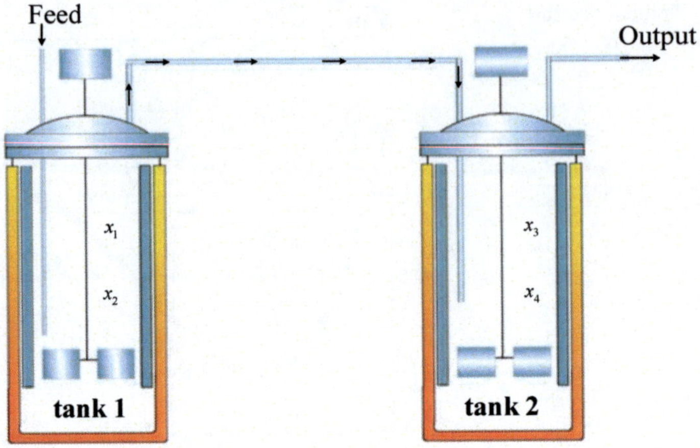


**Fig. 6.13** An illustration of the two-stage CSTRs

### *6.5.3 Optimal Control of a Chemical Reaction*

Continuously stirred tank reactors (CSTRs) are widely used in chemical engineering. An optimal control problem of a system with two-stage CSTRs is considered in this example, as illustrated in Fig. 6.13. A first-order irreversible chemical reaction occurs in each tank. And the system is governed by the following system equation

$$\begin{cases} \dot{x}_1(t) = 0.5 - x_1(t) - R_1 \\ \dot{x}_2(t) = -2[x_2(t) + 0.25] - u_1(t)[x_2(t) + 0.25] + R_1 \\ \dot{x}_3(t) = x_1(t - \theta) - x_3(t) - R_2 + 0.25 \\ \dot{x}_4(t) = x_2(t - \theta) - 2x_4(t) - u_2(t)[x_4(t) + 0.25] + R_2 - 0.25 \end{cases} \tag{6.84}$$

where variables $x_1$ and $x_3$, denotes the normalized concentration in tanks 1 and 2, respectively; and variable $x_2$ and $x_4$, represents the normalized temperatures in tanks 1 and 2, respectively; $u_1$ and $u_2$ are control variables that are related to the cooling jacket in tanks 1 and 2, respectively. The symbol $\theta$ represents the time delay in state variables and varies from 5 to 40% of the residence time in a single tank, i.e., $\theta \in [0.05, 0.4]$. The reaction term in each reactor tank is given as

$$R_1 = [x_1(t) + 0.5] \exp\left(\frac{25x_2(t)}{x_2(t) + 2}\right) \tag{6.85}$$

$$R_2 = [x_3(t) + 0.25] \exp\left(\frac{25x_4(t)}{x_4(t) + 2}\right) \tag{6.86}$$

If one denotes the state space as $\mathbf{x} = \left[x_1\ x_2\ x_3\ x_4\right]^{\mathrm{T}}$, the initial condition of the system is denoted as

$$\mathbf{x}(t) = \left[0.15\ -0.03\ 0.1\ 0\right]^{\mathrm{T}},\ \text{for}\ t \in [-\theta, 0] \tag{6.87}$$

And the goal is to minimize the following cost functional

$$J = \int_0^2 \left(0.1u_1^2 + 0.1u_2^2 + \mathbf{x}^{\mathrm{T}}\mathbf{x}\right)\mathrm{d}t \tag{6.88}$$

Then we consider the following two cases.

(1) Case I: system with no constraints

First, we consider $\theta = 0.4$. If we set $Q = 10$ and $N = 10$ in the SPM, and set $\varepsilon = 10^{-10}$ in the SCvx process, converged solutions can be obtained within 0.289932 s after 9 iterations. The state and control variables are plotted in Fig. 6.14 and Fig. 6.15, respectively.

The problem is also studied by the iterative dynamic programming (IDP) method in [23]. The computational results obtained by the IDP method and the SPM are reported in Table 6.4. It is seen that the results obtained by SPM have smaller cost functional.

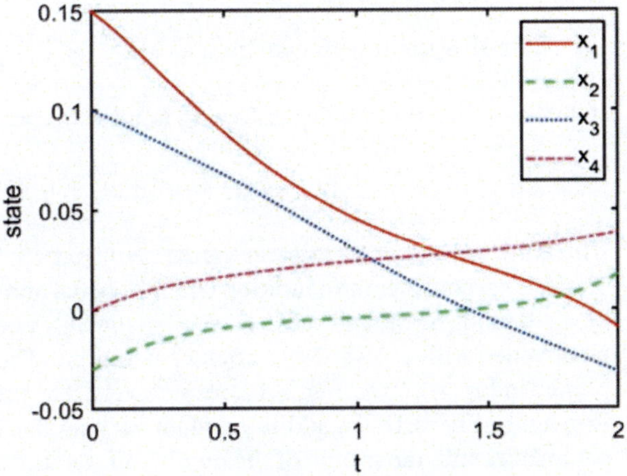


**Fig. 6.14** Numerical solutions of state variables

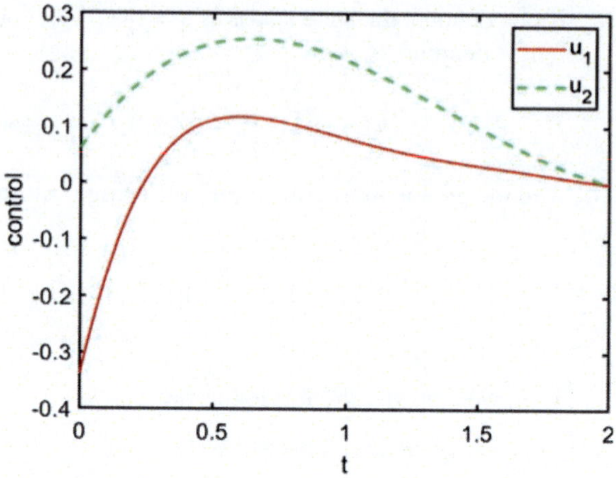


**Fig. 6.15** Numerical solutions of control variables

**Table 6.4** Numerical performances of different numerical methods

| State-delay $\theta$ | IDP method [23] | | SPM ($Q = 40$, $N = 10$) | |
|---|---|---|---|---|
| | Iterations | $J$ | Iterations | $J$ |
| 0.05 | 21 | 0.02297 | 9 | 0.02284919 |
| 0.10 | 21 | 0.02328 | 9 | 0.02316377 |
| 0.20 | 21 | 0.02372 | 9 | 0.02372899 |
| 0.40 | 27 | 0.02476 | 9 | 0.02461287 |

(2) Case II: system with constraints

In this case, we impose the following nonlinear constraints on control variables

$$h(t) = u_1^2 + u_2^2 - \bar{u} \leq 0 \tag{6.89}$$

where the parameter $\bar{u}$ varies in the range of [0.048, 0.08].

First, we consider the condition where $\theta = 0.4$ and $\bar{u} = 0.06$. If we set $Q = 10$ and $N = 10$ in the SPM, and set $\varepsilon = 10^{-10}$ in the SCvx process, converged solutions can be obtained within 2.389199 s after 22 iterations. The state and control variables are plotted in Fig. 6.16 and Fig. 6.17, respectively. The profile of constraint in Eq. (6.89), is given in Fig. 6.18. In addition, when varying the value of $\bar{u}$ while keeping other parameters still, the profile of the number of iterations is given in Fig. 6.19. It is seen that the number of iterations goes down as $\bar{u}$ increases. Actually, a lower $\bar{u}$ will result in a higher nonlinearity, leading the problem much harder to converge.

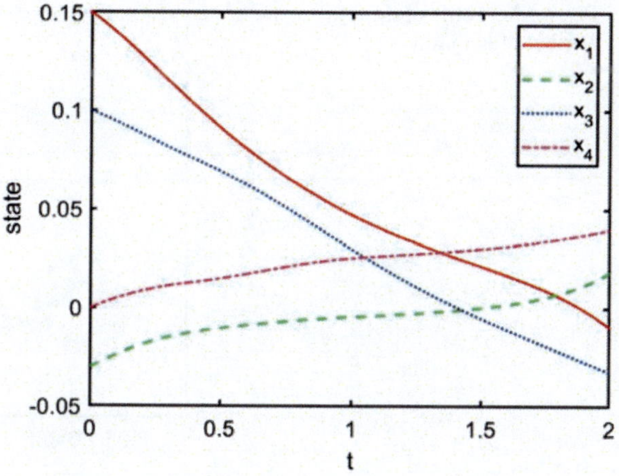


**Fig. 6.16** Numerical solutions of state variables

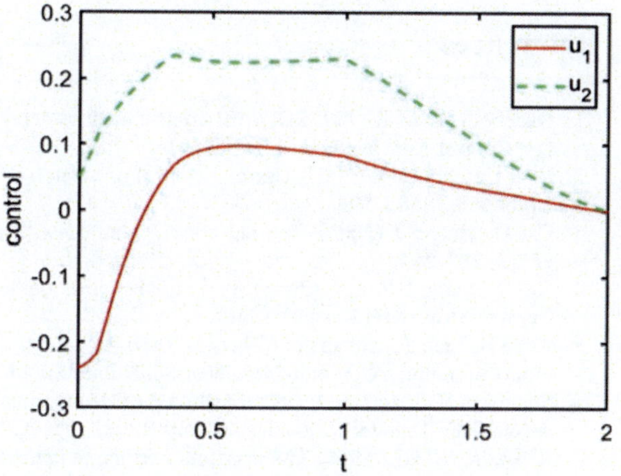


**Fig. 6.17** Numerical solutions of control variables

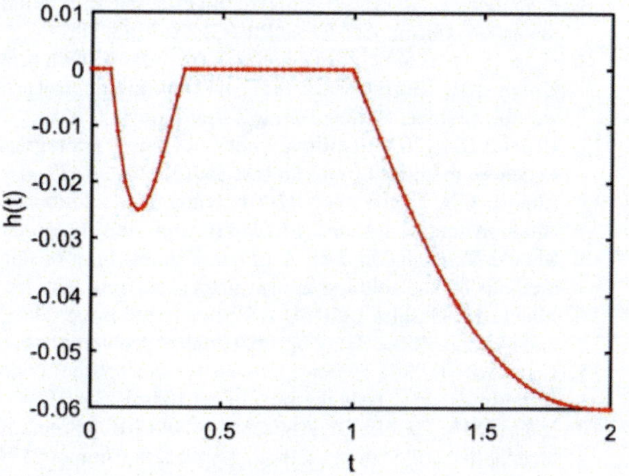


**Fig. 6.18** Profile of the constraint

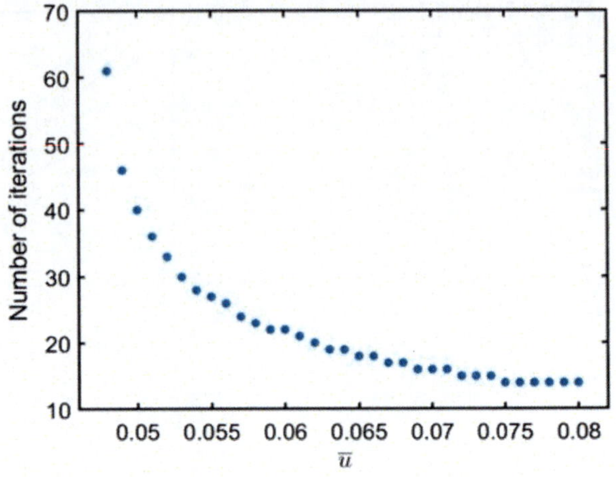


**Fig. 6.19** Profile of the number of iterations

## References


1. Dadebo S, Luus R (1992) Optimal control of time-delay systems by dynamic programming. Opt Control Appl Methods 13(1):29–41
2. Livk I, Rehbock V (2007) Optimal control of a batch crystallization process. J Indus Manag Optimiz 3(3):585–596
3. Cai G, Huang J (2002) Optimal control method with time delay in control. J Sound Vib 251(3):383–394
4. Cai G, Huang J, Yang S (2003) An optimal control method for linear systems with time delay. Comput Struct 81(15):1539–1546
5. Khan H, Liao S, Mohapatra RN et al (2009) An analytical solution for a nonlinear time-delay model in biology. Commun Nonlinear Sci Numer Simul 14(7):3141–3148
6. Shamsara E, Shamsara J, Afsharnezhad Z (2016) Optimal control therapy and vaccination for special HIV-1 model with delay. Theory Biosci 135(4):217–230
7. Kharatishvili GL (1961) The maximum principle in the theory of optimal process with time-lags. Dokl Akad Nauk SSSR 136:39–42
8. Kharatishvili GL (1967) A maximum principle in external problems with delays, mathematical theory on control. Academic Press, New York, NY
9. Halanay A (1968) Optimal controls for systems with time lag. SIAM J Control 6(2):215–234
10. Göllmann L, Kern D, Maurer H (2010) Optimal control problems with delays in state and control variables subject to mixed control–state constraints. Opt Control Appl Methods 30(4):341–365
11. Hoseini SM (2020) Optimal control of linear pantograph-type delay systems via composite Legendre method. J Franklin Inst 357(9):5402–5427
12. Elnagar GN, Kazemi MA (2001) Numerical solution of time-delayed functional differential equation control systems. J Comput Appl Math 130(1–2):75–90
13. Maleki M, Hadi-Vencheh A (2010) Combination of non-classical pseudospectral and direct methods for the solution of brachistochrone problem. Int J Comput Math 87(8):1847–1856
14. Maleki M, Hashim I (2014) Adaptive pseudospectral methods for solving constrained linear and nonlinear time-delay optimal control problems. J Franklin Inst 351(2):811–839
15. Perng MH (1986) Direct approach for the optimal control of linear time-delay systems via shifted Legend re polynomials. Int J Control 43(6):1897–1904
16. Maleki MA, Hashim IA, Abbasbandy SB (2012) Solution of time-varying delay systems using an adaptive collocation method. Appl Math Comput 219(4):1434–1448

17. Wang X, Peng H, Zhang S et al (2017) A symplectic local pseudospectral method for solving nonlinear state-delayed optimal control problems with inequality constraints. Int J Robust Nonlinear Control 28(6):2097–2120
18. Wang X, Liu J, Dong X et al (2020) A symplectic pseudospectral method for constrained time-delayed optimal control problems and its application to biological control problems. Optimization. https://doi.org/10.1080/02331934.2020.1786568
19. Haddadi N, Ordokhani Y, Razzaghi M (2012) Optimal control of delay systems by using a hybrid functions approximation. J Optim Theory Appl 153(2):338–356
20. Khellat F (2009) Optimal control of linear time-delayed systems by linear Legendre multiwavelets. J Optim Theory Appl 143(1):107–121
21. Banks HT, Burns JA (1978) Hereditary control problems: numerical methods based on averaging approximations. SIAM J Control Optim 16(2):169–208
22. Bonalli R, Hérissé B, Trélat E (2017) Solving optimal control problems for delayed control-affine systems with quadratic cost by numerical continuation. In: 2017 American control conference, 24–26 May 2017. Seattle, WA, USA
23. Chen C, Sun D, Chang C (2000) Numerical solution of time-delayed optimal control problems by iterative dynamic programming. Opt Control Appl Methods 21(3):91–105

# Chapter 7
# Model Predictive Control: From Open-Loop to Closed-Loop

## 7.1 Introduction

The SPMs developed in Chaps. 4–6, can be applied to solve trajectory planning problems with different features. They are actually open-loop control techniques, where control commands are computed based on the given model under ideal conditions. However, there could be various perturbations during the real control procedure. For example, the actual initial condition would deviate from the used when designing the trajectory using optima control methods. Additionally, external disturbances such as wind and base motion could be involved. An effective idea to solve such issues is to incorporate the feedback, leading to the closed-loop technique.

Closed-loop technology is mainly used in the field of trajectory tracking, such as trajectory tracking of unmanned ground systems (UGS), unmanned aerial vehicles (UAV), and robots. Although the closed-loop technology has been extensively studied, it's still difficult to solve the following problems, (1) the nonlinear and coupling characteristics of the system; (2) constraints of input and output of the system; (3) the dynamic environment. The widely used tracking methods can be divided into 5 categories: geometric and kinematic tracking control methods [1], classic tracking control methods [2–6], tracking control methods based on dynamic state feedback (linear and nonlinear) [7–9], model prediction methods (MPC) [10–14], neural network-based, and fuzzy logic tracking method [15–19].

Geometric and kinematic tracking control methods are mainly used in the trajectory tracking of vehicles, it's easy to implement, but the dynamics of the system is hard to be considered. The methods based on geometric mainly include Pure-pursuit and Stanley methods. Pure-pursuit is a method of constructing a virtual point in front of a moving body, and the tracking of the system can be realized by tracking the virtual point. Stanley calculates the steering angle which should be corrected



X. Wang et al., *Symplectic Pseudospectral Methods for Optimal Control*, Intelligent Systems, Control and Automation: Science and Engineering 97, https://doi.org/10.1007/978-981-15-3438-6_7

by considering the longitudinal and lateral errors, it's difficult to balance the relationship between stability and tracking performance. The kinematics-based control method is to design a feedback controller based on the kinematics of the system, which can improve the tracking performance of the geometric controller. However, these methods don't consider the dynamics of the system, there is no guarantee of the applicability with high-speed mode.

Classical control methods mainly include PID control and sliding mode control (SMC). PID controller is simple and widely used in industry, it's usually triggered by the error between the actual and the expected response. The tracking controller using classic PID technology has good tracking performance, but the effect is very sensitive to the controller parameters. SMC is a mature nonlinear state feedback controller, the control obtained by using SMC has good tracking accuracy. These methods can effectively solve specific problems, but also have some shortcomings, (1) the tracking control performance is sensitive to the sampling period; (2) it's only robust on the sliding surface; (3) prior knowledge of disturbance and uncertainty are required.

The tracking control methods based on dynamic state feedback (linear and nonlinear) has better control performance than the geometric controller while considering system dynamics or kinematics. Recent years, nonlinear adaptive control technologies such as Inversion & Immersion (I&I) have also been applied to the field of trajectory tracking. Preliminary research shows that this method has robust closed-loop tracking performance, but the controller is sensitive to parameter uncertainty.

In addition, the linearization method is widely used to convert the original nonlinear system into a linear time-varying/time-invariant system, and the controller will be designed based on the linear control theory to achieve tracking. It can simplify the problem to obtain a linear model by using the first-order Taylor expansion of complex nonlinear systems, and there is a relatively mature linear control theory to solve it, which can achieve fast solution too. However, as this method ignores higher-order terms, the linearized model may be difficult to describe some characteristics of the original system when the tracking error or disturbance is large. With the continuous development of computer technology, some researchers try to use nonlinear theory and numerical calculation theory to solve this problem. Nonlinear MPC can achieve very accurate tracking, but the calculation efficiency will be affected so that it's difficult to achieve online tracking control.

The tracking methods based on neural networks and fuzzy logic have the ability to guide the controller to make quick decisions, they have adaptive and intelligent characteristic. Although adaptive/smart controllers can provide a robust solution, it should be guided by "expertise" or "experience". Therefore, these algorithms need to learn and train with a large amount of a priori information to obtain more "expert knowledge", so that it can make the correct response to the actual situation. However, as the formal stability proofs are hard to be provided, these methods have not been well applied to actual equipment or equipment.

Compared to the methods mentioned above, model predictive control (MPC) owns the capability to handle constraints. At each sampling instant, an optimization problem is required to be solved. It could be extremely time-consuming when the optimization problem is not well formulated or subjected to too many complex constraints. Hence, the bottleneck that limits the application of MPC in practice is its online computational efficiency. In this chapter, we will take the SPM for nonlinear optimal control problems developed in Chap. 5 as the core solver to construct the symplectic pseudospectral MPC. Only the implementation of MPC is provided, characters of the MPC such as stability are not considered herein.

## 7.2 The Basic Idea of MPC

Consider the controlled system subjected to the following system equations

$$\dot{\mathbf{x}} = \mathbf{f}(\mathbf{x}, \mathbf{u}, t), \ \mathbf{x}(t_s) = \mathbf{x}_s \tag{7.1}$$

where $\mathbf{x}(t) \in \mathbb{R}^d$ is the state variable, and its value at $t = t_s$ is denoted as $\mathbf{x}_s$; $\mathbf{u}(t) \in \mathbb{R}^p$ is the control variable. Supposing the sampling instants are $\{\tilde{t}_k\}_{k=1}^{\infty}$ with prescribed fixed sampling period $\delta T = \tilde{t}_{k+1} - \tilde{t}_k$. Additionally, we set $\tilde{t}_1 = t_s$.

It is assumed that all state variables can be measured (or estimated by filters). The basic idea of MPC is to solve an open-loop optimal control problem where the system equations and possible constraints are involved meanwhile the current state is taken as initial conditions. Denoted the current state at the $k$ th sampling instant as $\mathbf{X}_k^0$, the open-loop optimal control problem to be solved at this instant is formulated as follows:

$$\text{Problem }(\text{P}_\text{k}) \begin{cases} \min J = E\big(\mathbf{x}(\tilde{t}_k + T)\big) + \displaystyle\int_{\tilde{t}_k}^{\tilde{t}_k+T} L(\mathbf{x}, \mathbf{u}, t)\mathrm{d}t \\ s.t. \\ \dot{\mathbf{x}} = \mathbf{f}(\mathbf{x}, \mathbf{u}, t), k = 1, 2, 3, \ldots \\ \mathbf{x}(\tilde{t}_k) = \mathbf{X}_k^0 \\ \mathbf{h}(\mathbf{x}, \mathbf{u}, t) \leq \mathbf{0} \end{cases} \tag{7.2}$$

where $[\tilde{t}_k, \tilde{t}_k + T]$ is the current prediction window, and $T$ is called the length of prediction window; $E : \mathbb{R}^d \to \mathbb{R}$ and $L : \mathbb{R}^d \times \mathbb{R}^p \times \mathbb{R} \to \mathbb{R}$ are functions to describe the terminal term and integration term in the cost functional. Denote the control solutions of Problem ($\text{P}_\text{k}$) as $\mathbf{u}^*(t)(\tilde{t}_k \leq t \leq \tilde{t}_k + T)$, the first step in the control sequence, i.e., $\mathbf{u}^*(\tilde{t}_k)$ is inserted as the control input as this sampling instant. A general illustration of the MPC is given in Fig. 7.1.

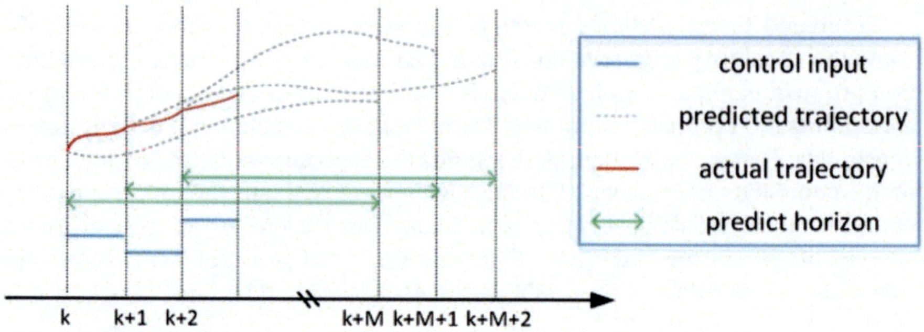


**Fig. 7.1** An illustration of the model predictive control

**Remark 7.1** In most scenarios, the lengths of prediction window at every sampling instant are fixed. However, such the setting is not fixed. For some special cases, the selection of the window length could be flexible. For example, to complete the transfer of spacecraft between different Halo orbits exactly at $t = T^*$ under low-thrust control, the terminal end of each prediction window is fixed at $t = T^*$ in [20]. □

## References

1. Mike M, Kenny L, Pascal S et al (2006) Stanley: the robot that won the darpa grand challenge. J Field Robot 23(9):661–692
2. Pan Y, Li X, Yu H (2018) Efficient PID tracking control of robotic manipulators driven by compliant actuators. IEEE Trans Control Syst Technol 27(2):915–922
3. Zhu R, Sun D, Zhou Z (2007) Integrated design of trajectory planning and control for micro air vehicles. Mechatronics 17(4):245–253
4. Normey-Rico JE, Alcalá I, Gómez-Ortega J et al (2001) Mobile robot path tracking using a robust PID controller. Control Eng Practice 9(11):1209–1214
5. Matraji I, Al-Durra A, Haryono A et al (2018) Trajectory tracking control of skid-steered mobile robot based on adaptive second order sliding mode control. Control Eng Practice 72:167–176
6. Muñoz F, Espinoza ES, González-Hernández I et al (2018) Robust trajectory tracking for unmanned aircraft systems using a nonsingular terminal modified super-twisting sliding mode controller. J Intell Robot Syst (1):1–18
7. Ajjanaromvat N, Parnichkun M (2018) Trajectory tracking using online learning LQR with adaptive learning control of a leg-exoskeleton for disorder gait rehabilitation. Mechatronics 51:85–96
8. Snider JM (2009) Automatic Steering methods for autonomous automobile path tracking. Robotics Institute, Carnegie Mellon University, Pittsburgh
9. Tagne G, Talj R, Charara A (2016) Design and comparison of robust nonlinear controllers for the lateral dynamics of intelligent vehicles. IEEE Trans Intell Transp Syst 17(3):796–809
10. Borrelli F, Falcone P, Keviczky T et al (2005) MPC-based approach to active steering for autonomous vehicle systems. Int J Veh Auton Syst 3(2/3/4):265
11. Falcone P, Borrelli F, Asgari J et al (2007) Predictive active steering control for autonomous vehicle systems. IEEE Trans Control Syst Technol 15(3):566–580

12. Bahadorian M, Eaton R, Hesketh T et al (2014) Robust time-varying model predictive control with application to mobile robot unmanned path tracking. IFAC Proc Vol 47(3):4849–4854
13. Gutjahr B, Gröll L, Werling M (2017) Lateral vehicle trajectory optimization using constrained linear time-varying MPC. IEEE Trans Intell Transp Syst 18:1586–1595
14. Li Z et al (2017) Trajectory-tracking control of mobile robot systems incorporating neural-dynamic optimized model predictive approach. IEEE Trans Syst Man Cybern Syst 46(6):740–749
15. Fukao T (2000) Inverse optimal tracking control of a nonholonomic mobile robot. IEEE Trans Robot Autom 16(5):609–615
16. Shirzadeh M, Asl HJ, Amirkhani A, et al (2017) Vision-based control of a quadrotor utilizing artificial neural networks for tracking of moving targets. Eng Appl Artif Intell 58:34–48
17. Jiang P, Unbehauen R (2002) Iterative learning neural network control for nonlinear system trajectory tracking. Neurocomputing 48(1):141–153
18. Moreno-Valenzuela J, Aguilar-Avelar C, Puga-Guzmán SA et al (2016) Adaptive neural network control for the trajectory tracking of the furuta pendulum. IEEE Trans Cybern 46(12):3439
19. Amer NH, Zamzuri H, Hudha K et al (2017) Modelling and control strategies in path tracking control for autonomous ground vehicles: a review of state of the art and challenges. J Intell Robot Syst 86(2):1–30
20. Peng H, Jiang X, Chen B (2014) Optimal nonlinear feedback control of spacecraft rendezvous with finite low thrust between libration orbits. Nonlinear Dyn 76(2):1611–1632

# Chapter 8
# Optimal Maneuver for Spacecraft

## 8.1 Introduction

Since the birth of optimal control theory, astronautical engineering is always its most important application scenario. In this chapter, we solve some optimal maneuver problem by the symplectic pseudospectral methods developed in Part II.

## 8.2 Optimal Rendezvous of Spacecraft Around the Earth

Consider the spacecraft subjected to the central gravity field of the Earth [11]. In a coordinate system that is rotating along a circular orbit around the Earth at a constant angular velocity, the motion equations of the spacecraft can be written into the following non-dimensional formulation.

$$\ddot{x} - 2\dot{y} + (1 + x)\left(\frac{1}{r^3} - 1\right) = u_x \tag{8.1}$$

$$\ddot{y} + 2\dot{x} + y\left(\frac{1}{r^3} - 1\right) = u_y \tag{8.2}$$

$$\ddot{z} + \frac{1}{r^3} z = u_z \tag{8.3}$$

where



X. Wang et al., *Symplectic Pseudospectral Methods for Optimal Control*, Intelligent Systems, Control and Automation: Science and Engineering 97,
https://doi.org/10.1007/978-981-15-3438-6_8

$$r^2 = (1 + x)^2 + y^2 + z^2 \tag{8.4}$$

The variable $x$, $y$, and $z$ represent the coordinates of radial, tangential, and normal displacements from the origin of the rotating frame, respectively. And $u_x$, $u_y$, $u_z$ denote the control forces applied in the $x$, $y$, and $z$ directions, respectively. In the rendezvous process, the initial time is $t_s = 0$. The spacecraft is transferred from the initial state to the goal state at the terminal time $t_f = 1$. And the boundary conditions are set as follows:

$$\begin{cases} x(0) = y(0) = y(0) = 0.3 \\ \dot{x}(0) = \dot{y}(0) = \dot{z}(0) = 0.2 \\ x(1) = y(1) = y(1) = 0 \\ \dot{x}(1) = \dot{y}(1) = \dot{z}(1) = 0 \end{cases} \tag{8.5}$$

No extra path or control constraints are considered in this example. We aim at designing an energy-optimal transfer trajectory, where the cost functional is

$$J = \frac{1}{2}\int_0^1 (u_x^2 + u_y^2 + u_z^2)dt \tag{8.6}$$

The SPM for unconstrained problems together with the *hp* mesh refinement techniques developed in Chap. 4, is used for numerical solving. In the mesh refinement process, we set $\mu = 1.0e-8$, $R = 1.1$, $A = 1$, $B = 1.2$, and $iterMax = 20$. As for the initial mesh, the time domain is divided into four regular sub-intervals and a 4-order LGL pseudospectral approximation is applied in each sub-interval. Converged solutions, as illustrated in Figs. 8.1 and 8.2, can be obtained within two

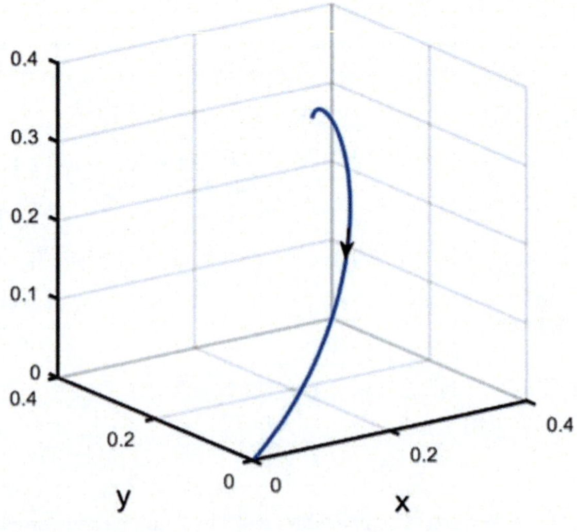


**Fig. 8.1** Position of the spacecraft

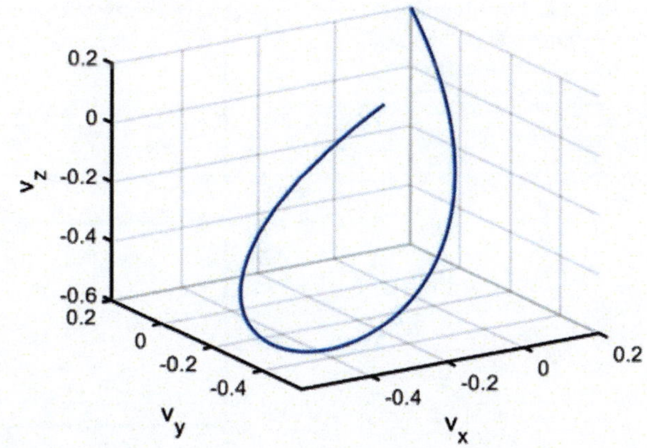


**Fig. 8.2** Velocity phase of the spacecraft

mesh refinement iterations. And the profiles of control input are plotted in Fig. 8.3.

In addition, distributions of mesh end and LGL node in each mesh refinement iteration are given in Figs. 8.4 and 8.5. In Table 8.1, the history of parameters in each iteration is summarized. It is seen that the *p* mesh refinement process is applied within the time domain [0.75, 1], which implies the relative curvature of trajectory of the each state variable is small in this region.

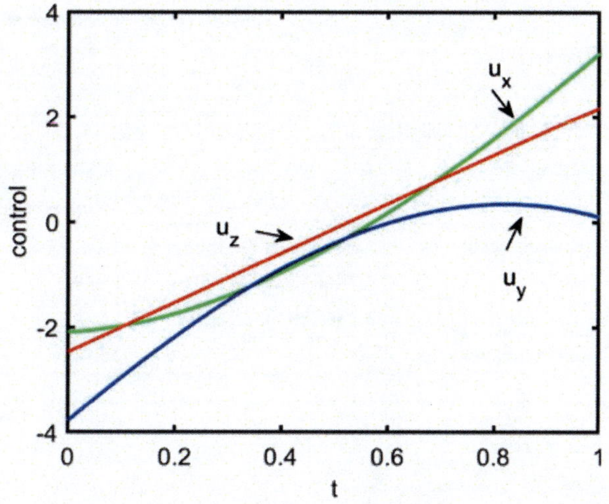


**Fig. 8.3** Profiles of control inputs

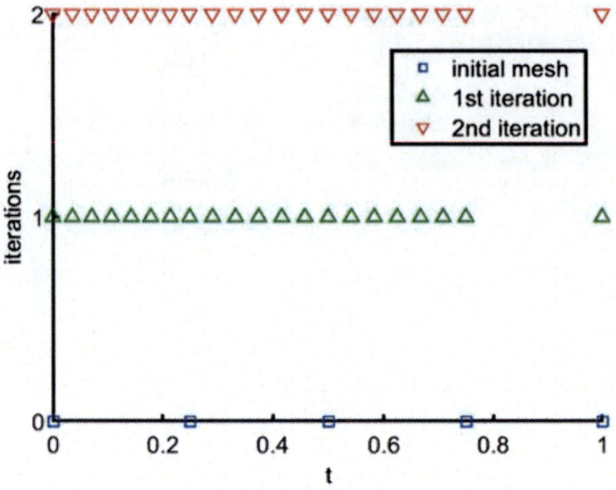


**Fig. 8.4** Distribution of mesh interval in the mesh refinement process

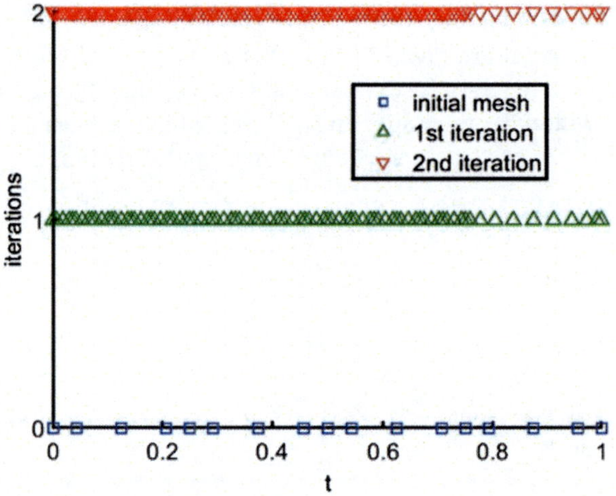


**Fig. 8.5** Distribution of LGL nodes in the mesh refinement process

**Table 8.1** Distribution of mesh interval in the mesh refinement process

| Iterations | $J$ | CPU time/s | | |
|---|---|---|---|---|
| | | Remeshing | Solving | Total |
| Initial mesh | **3.386**065790014565 | – | 0.022 | 0.022 |
| 1 | **3.386820**013724094 | 0.003 | 0.303 | 0.306 |
| 2 | **3.386820401**752699 | 0.006 | 1.051 | 1.058 |
| 3(final check) | – | 0.006 | – | 0.006 |

## 8.3 Optimal Transfer Between Halo Orbits

Consider the transfer of a spacecraft between the Halo orbits around the $L_2$ libration point of the Sun-Earth system (as illustrated in Fig. 8.6) [12, 13]. The motion of the spacecraft is model by the following nondimensional equations [14]:

$$\ddot{x} - 2\dot{y} - x = -\frac{(1-\mu)(x+1+1/\gamma)}{\gamma^3 d_1^3} - \frac{\mu(x+1)}{\gamma^3 d_2^3} + \frac{1-\mu+\gamma}{\gamma} + u_x \quad (8.7)$$

$$\ddot{y} + 2\dot{x} - y = -\frac{(1-\mu)y}{\gamma^3 d_1^3} - \frac{\mu y}{\gamma^3 d_2^3} + u_y \quad (8.8)$$

$$\ddot{z} = -\frac{(1-\mu)z}{\gamma^3 d_1^3} - \frac{\mu z}{\gamma^3 d_2^3} + u_z \quad (8.9)$$

where

$$d_1 = \sqrt{(x+1+1/\gamma)^2 + y^2 + z^2} \quad (8.10)$$

$$d_2 = \sqrt{(x+1)^2 + y^2 + z^2} \quad (8.11)$$

The control inputs $u_i (i = x, y, z)$ satisfy the following nonlinear constraints

$$h(t) = u_x^2 + u_y^2 + u_z^2 - \bar{u}^2 \leq 0 \quad (8.12)$$

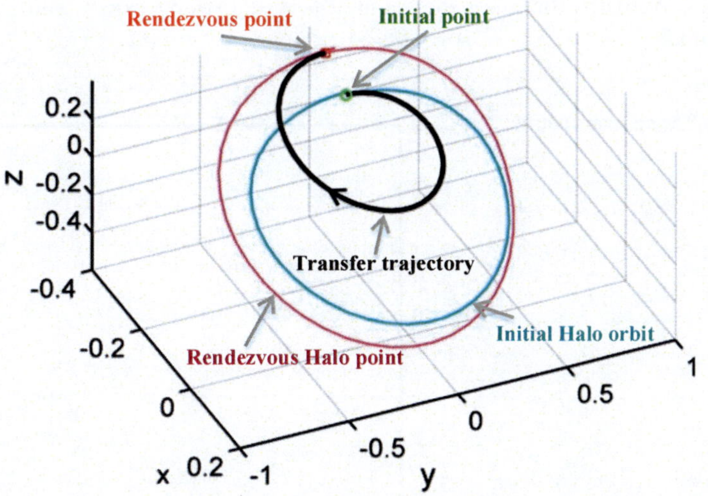


**Fig. 8.6** The transfer trajectory between two Halo orbits

The initial and the final desired conditions are defined as follows:

$$\begin{cases} x(0) = -0.252566564360231, x(t_f) = -0.334262861958610 \\ y(0) = 0, y(t_f) = 0 \\ z(0) = 0.273165147711868, z(t_f) = 0.366333802467158 \\ \dot{x}(0) = 0, \dot{x}(t_f) = 0 \\ \dot{y}(0) = 1.254563849725983, \dot{y}(t_f) = 1.498981257524648 \\ \dot{z}(0) = 0, \dot{z}(t_f) = 0 \end{cases} \tag{8.13}$$

One aims to design a transfer trajectory satisfying Eqs. (8.7)– (8.13), meanwhile minimizing the following cost functional

$$J = \int_0^{t_f} \left(u_x^2 + u_y^2 + u_z^2\right) dt \tag{8.14}$$

Several parameters in the above equations are set as follows: $\mu = 3.04036e - 6$, $\gamma = 1.00782e - 2$, and $t_f = 1.63420$. The parameter $\bar{u}$ in Eq. (8.12), is the upper limit of control, and it varies from 2.5 to 3.5. This example is solved by the SPM for constraints with inequality constraints developed in Chap. 5, and the convergent tolerance is set as $\varepsilon = 1 \times 10^{-8}$. To implement the SPM, the time domain is divided into $Q = 10$ sub-intervals and a 10-order approximation polynomial is applied in each sub-interval. When one set $\bar{u} = 2.5$, converged solutions as illustrated in Figs. 8.6 and 8.7, are obtained after 49 iterations. Profiles of the constraint defined in Eq. (8.12), is also given in Fig. 8.8. It is seen that this constraint is strictly satisfied within the while control horizon.

The variation of the number of iterations along with the value of $\bar{u}$ is given in Fig. 8.9. It is seen that the number of iterations decreases quickly as the value of $\bar{u}$ increases. Actually, the lower the value of $\bar{u}$ is, the stronger the nonlinearity of the problem is.

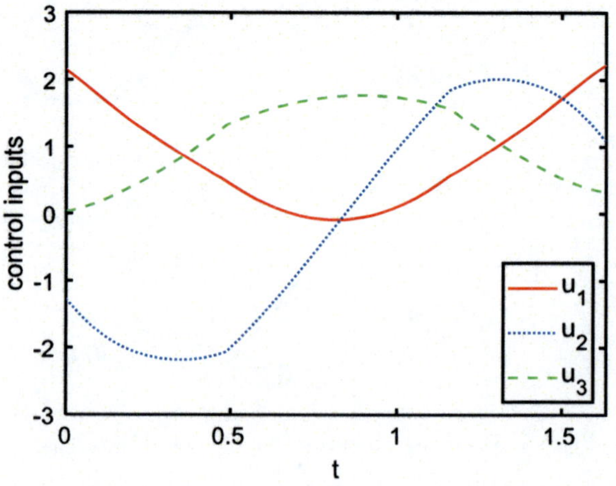


**Fig. 8.7** The optimal control inputs

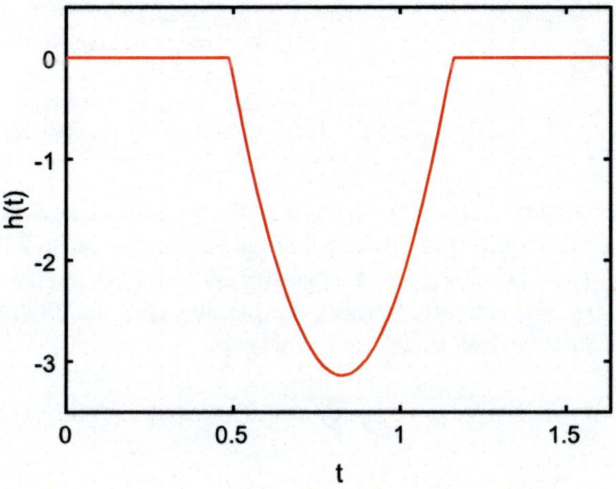


**Fig. 8.8** Profile of the constraint in Eq. (8.12)

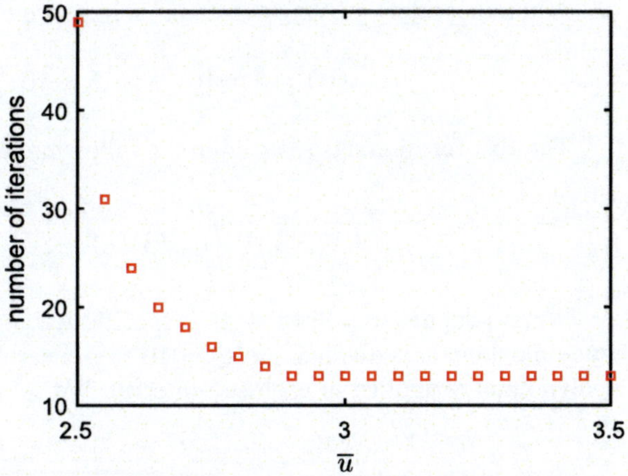


**Fig. 8.9** The transfer trajectory between two Halo orbits

## 8.4 Time-Energy Hybrid Optimal Attitude Maneuver of Spacecraft

Examples in Sects. 8.2 and 8.3 are both with fixed terminal time. In this section, the time-energy optimal attitude maneuver of rigid asymmetric spacecraft is considered. The Euler's equations for the angular velocities $x_1(t)$, $x_2(t)$ and $x_3(t)$ are as follows (unit: rad/s) [15, 16]:

$$\dot{x}_1(t) = -\frac{I_3 - I_2}{I_1} x_2(t) x_3(t) + \frac{u_1(t)}{I_1} \tag{8.15}$$

$$\dot{x}_2(t) = -\frac{I_1 - I_3}{I_2} x_3(t) x_1(t) + \frac{u_2(t)}{I_2} \tag{8.16}$$

$$\dot{x}_3(t) = -\frac{I_2 - I_1}{I_3} x_1(t) x_2(t) + \frac{u_3(t)}{I_3} \tag{8.17}$$

where $u_1(t)$, $u_2(t)$, and $u_3(t)$ are the control torques (unit: $\mathrm{N \cdot m}$), and the momentum of rotation along three principal axes are $I_1 = 86.24\mathrm{kg \cdot m^2}$, $I_2 = 85.07\mathrm{kg \cdot m^2}$, and $I_3 = 113.59\mathrm{kg \cdot m^2}$, respectively. The initial and the final times are denoted as $t_s$ and $t_f$, respectively. In this example, we set $t_s = 0$ while $t_f$ is free. The initial and the final desired states are as follows:

$$x_1(t_s) = 0.01, \; x_2(t_s) = 0.001, \; x_3(t_s) = 0.001 \tag{8.18}$$

$$x_1(t_f) = x_2(t_f) = x_3(t_f) = 0 \tag{8.19}$$

A nonlinear time-varying constraint is imposed on the state $x_1(t)$ as follows:

$$x_1(t) \leq 5 \times 10^{-6} t^2 - 5 \times 10^{-4} t + 0.016 \tag{8.20}$$

The cost functional is selected in the following fashion

$$J = \omega t_f + \int_{t_s}^{t_f} \left( u_1^2(t) + u_2^2(t) + u_3^2(t) \right) \mathrm{d}t \tag{8.21}$$

When adopting the SPM with unspecific terminal time to solve this example, the time domain is divided into $Q = 10$ sub-interval and a 20-order approximation polynomial is applied in each sub-interval. We set the weight on terminal time as

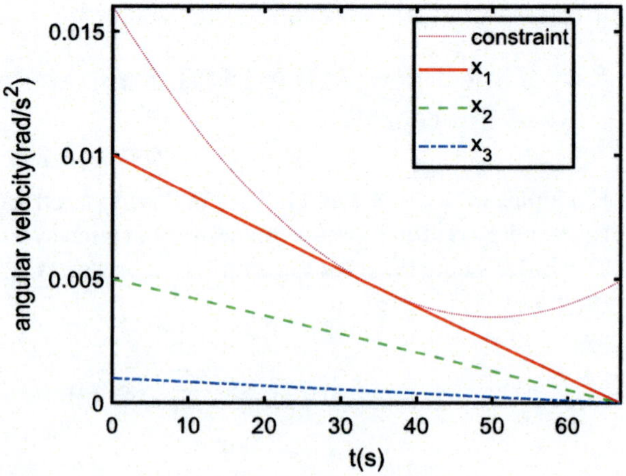


**Fig. 8.10** Profiles of state variables

$\omega = 1 \times 10^{-4}$, and two initial guesses on the terminal time are set as $t_f^{(1)} = 90$ and $t_f^{(2)} = 80$. Other solution parameters are set as $\varepsilon = 10^{-6}$, $\varepsilon' = 10^{-4}$ and $\tau = 10^{-3}$. Then converged solutions are obtained by seven iterations on the terminal time, and the final optimal maneuver time is $t_f^{*} = 66.9909$. The optimal state trajectories and optimal control torques are plotted in Fig. 8.10 and Fig. 8.11, respectively.

The iteration history of the terminal time is given in Fig. 8.12. It is seen that the terminal time converges to its optimal value quickly.

**Fig. 8.11** Profiles of control variables

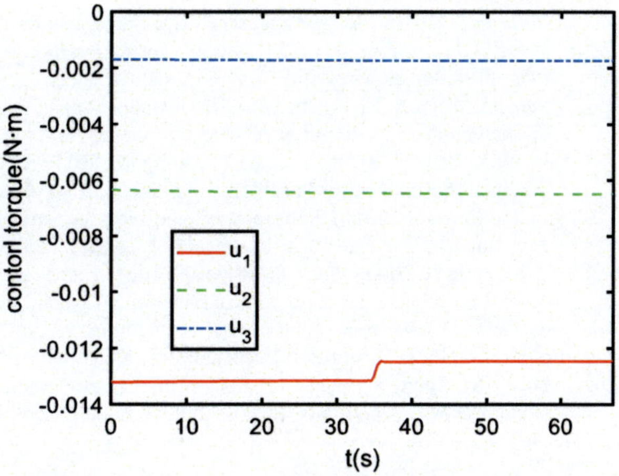


**Fig. 8.12** Iteration history of the terminal time

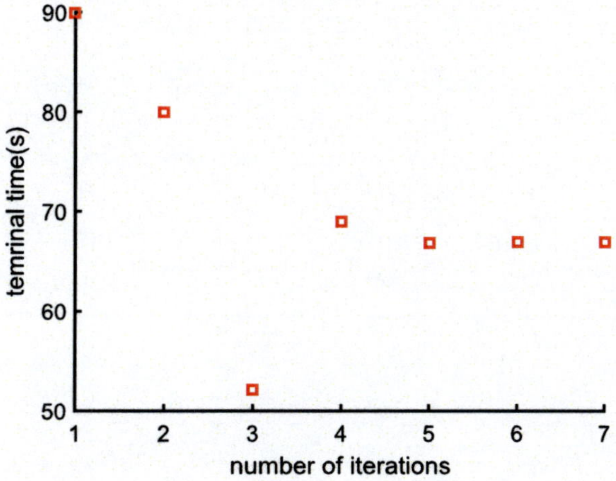

## References

1. Ma X, Fang J, Ning X (2013) An overview of the autonomous navigation for a gravity-assist interplanetary spacecraft. Progress Aerosp Sci 63(Nov):56–66
2. Shirazi A, Ceberio J, Lozano JA (2019) Spacecraft trajectory optimization: a review of models, objectives, approaches and solutions. Progress Aerosp Sci 102(Oct):76–98
3. Chai R, Savvaris A, Tsourdos A, et al (2019) A review of optimization techniques in spacecraft flight trajectory design. Progress Aerosp Sci 109:Article 100543
4. Ovchinnikov MY, Roldugin DS (2019) A survey on active magnetic attitude control algorithms for small satellites. Progress Aerosp Sci 109:Article 100546
5. Fehse W (2003) Automated rendezvous and docking of spacecraft. Cambridge University Press, Cambridge
6. Ben-Asher JZ (2010) Optimal control theory with aerospace applications. American Institute of Aeronautics and Astronautics, Inc
7. Tewari A (2011) Advanced control of aircraft, spacecraft and rockets. Wiley
8. Longuski JM, Guzmán JJ, Prussing JE (2014) Optimal control with aerospace applications. Springer, New York
9. Mazzini L (2016) Flexible spacecraft dynamics, control and guidance. Springer, Cham
10. Wang D, Wu B, Poh EK (2017) Satellite formation flying: relative dynamics, formation design, fuel optimal maneuvers and formation maintenance. Springer, Singapore
11. Peng H, Wang X, Li M, Chen B (2017) An hp symplectic pseudospectral method for nonlinear optimal control. Commun Nonlinear Sci Numer Simul 42:623–644
12. Wang X, Peng H, Zhang S, Chen B, Zhong W (2017) A symplectic pseudospectral method for nonlinear optimal control problems with inequality constraints. ISA Trans 68:335–352
13. Li M, Peng H (2016) Solutions of nonlinear constrained optimal control problems using quasilinearization and variational pseudospectral methods. ISA Trans 62:177–192
14. Li M, Peng H, Zhong W (2015) A symplectic sequence iteration approach for nonlinear optimal control problems with state-control constraints. J Franklin Inst 352(6):2381–2406
15. Yousef ET, Lakestani M (2015) Direct solution of nonlinear constrained quadratic optimal control problems using B-spline functions. Kybernetika 51(1):81–98
16. Shang M, Zhang T, Song J (2013) Optimal torque control constraint rigid spacecraft attitude maneuver with constraints. In: The 25th Chinese control and decision conference (CCDC)

# Chapter 9
# Optimal Path Planning of UGS

## 9.1 Introduction

Nowadays, unmanned ground systems (UGS) have broad prospects in many fields, involving transportation, military, industrial, agricultural, and space exploration, etc., which has a deep impact on the national economy, national defense security, and social progress. According to incomplete statistics from authoritative departments, there are more than 300 types of ground unmanned systems, including more than 200 types applied in the military field. UGSs are autonomous decision-making systems that should consider the integration of vehicles, environment, and many technologies such as automatic control, artificial intelligence, sensor technology, data structure, navigation, and positioning. To realize stable and safe autonomous movement or complete specific tasks, a feasible path between the current and the final desired target is required. Therefore, path planning is one of the core technologies of UGS.

## 9.2 An Overview of the Methods of UGS Path Planning

Path planning techniques for UGS are generally categorized into four aspects, i.e., node-based search algorithm, artificial potential field method, intelligent algorithm, and optimal control method.

Node-based search algorithm mainly searches several nodes in the feasible region by dispersing the map environment into small regions and then connects the nodes to generate feasible paths [1–3]. It can be divided into two categories: node search algorithms based on graph theory and cell decomposition methods. This kind of algorithms is of high efficiency, but it usually does not consider the dynamic or kinematic equations [4–7].



X. Wang et al., *Symplectic Pseudospectral Methods for Optimal Control*, Intelligent Systems, Control and Automation: Science and Engineering 97, https://doi.org/10.1007/978-981-15-3438-6_9

Artificial potential field method adopts the concept of potential field in physics. The target point provides gravitational field to the optimized object, the obstacle provides repulsion field, and the path is generated by using the mechanical theory [8–10]. It can be highly efficient, but it's difficult to ensure that the optimized object can reach the destination accurately. Moreover, it ignores mechanical constraints such as the minimum turning radius of UGS.

Intelligent algorithm is mainly based on the existing path planning data, and it adopts artificial intelligence method to train the path planning model [11]. It mainly includes reinforcement learning, SVM, and so on. This kind of algorithms has strong self-learning ability and high intelligence, but it relies heavily on data or sampling, which requires a lot of scientific and reasonable training [12]. In addition, it is hard to guarantee the satisfaction of various kinds of constraints.

Optimal control method formulateds the path planning problem into a dynamic optimization problem where various kinds of constraints can be treated under a uniform framework [13–15]. Specifically, the transfer time and energy consumption can be considered synthetically in the performance index. The advantages of this method are twofold. First, the obtained trajectory considers kinematics/dynamics of UGSs, control inputs can be obtained simultaneously, as well as the trajectory. Hence, the obtained path can be better applied to engineering practice. Secondly, all kinds of constraints can be considered comprehensively, a unified framework can be used. However, it is computationally complicated where the improvement of efficiency is a prerequisite for the wide application of this method.

## 9.3 Problem Description

### 9.3.1 Cost Functional

The cost functional can be generally selected in time optimal/energy optimal/hybrid time-energy optimal form.

If it's required to maintain the smoothness of the movement with a fast transfer, the "time-energy optimal" objective function can be formulated as follows:

$$J = k(t_f - t_0) + \frac{1}{2}\int_{t_0}^{t_f} (\mathbf{U}^{\mathrm{T}}\mathbf{R}\mathbf{U})\mathrm{d}t \tag{9.1}$$

If the transfer time is not the major concern, an "energy optimal" objective function can be selected as follows:

$$J = \frac{1}{2}\int_{t_0}^{t_f} (\mathbf{U}^{\mathrm{T}}\mathbf{R}\mathbf{U})\mathrm{d}t \tag{9.2}$$

### 9.3.2 Constraints

(1) Motion equations

System modeling is the prerequisite to conduct path planning by the optimal control method, where accurate motion equations are required. The motion equations of UGSs can generally be divided into two categories, i.e., kinematic model and dynamic model. The kinematic models are more prominent at low speeds, while dynamic models can better describe the system with high speeds.

Kinematic models are relatively simple, however, they can't meet the accuracy and stability of lateral control when the speed is high. Compared with kinematic models, dynamic models can describe the behavior of UGS more accurately. However, many parameters are involved in dynamic models, which increases the difficulty of application. The selection between dynamics and kinematic models depends on the application scenario. Motion equations can be generally formulated as a set of ordinary equations as follows:

$$\dot{\mathbf{X}} = \mathbf{f}(\mathbf{X}, \mathbf{U}, t) \tag{9.3}$$

where the $\mathbf{X}$ is the state variable list, $\mathbf{U}$ is the control variable list.

(2) State and control constraints

Subject to the constraints of the system's structure and the safety requirements, the state variables and control variables should be constrained within a certain range. The constraints of state variables and control variables can be expressed as follows:

$$\tilde{\mathbf{h}}(\mathbf{X}, \mathbf{U}, t) \leq \mathbf{0} \tag{9.4}$$

(3) Collison constraints

Consider $n$ obstacles in the operational environment. For the $i$th $(i = 1, 2, \ldots, n)$ obstacle, the collision-free conditions can be expressed in the following fashion

$$1 - \left(\frac{x_C - x_{\mathrm{obs,i}}}{a_i + r + r_s}\right)^{2p_i} - \left(\frac{y_C - y_{\mathrm{obs,i}}}{b_i + r + r_s}\right)^{2p_i} \leq 0 \tag{9.5}$$

The UGS will be described as a circle with the radius $r$, and $x_C$ and $y_C$ are the center location of the UGS. Variables $x_{\mathrm{obs,i}}$ and $y_{\mathrm{obs,i}}$ are the center location of the $i$ th obstacle, $a_i$ and $b_i$ are its width and height, $p_i \in \mathbb{R}^+$ describes its shape [16, 17]. And $r_s$ is the buffer distance.

Inequality constraints in Eqs. (9.4) and (9.5), can be denoted as follows:

$$\mathbf{h}(\mathbf{X}, \mathbf{U}, t) \leq \mathbf{0} \tag{9.6}$$

### 9.3.3 Optimal Path Planning Model of UGS

According to the Eqs. (9.1)–(9.3) and (9.6), the optimal path planning model of UGSs can be denoted as follows:

$$\begin{cases} \min J \\ s.t. \\ \mathbf{X}(t_0) = \mathbf{X}_s, \mathbf{X}(t_f) = \mathbf{X}_f \\ \dot{\mathbf{X}} = \mathbf{f}(\mathbf{X}, \mathbf{U}, t) \\ \mathbf{h}(\mathbf{X}, \mathbf{U}, t) \le \mathbf{0} \end{cases} \tag{9.7}$$

where $t_0$ and $t_f$ are the departure time and arrival time, respectively; $\mathbf{X}_s$ and $\mathbf{X}_f$ are the corresponding boundary conditions; $\mathbf{R}$ is a positive definite weight matrix. Using the proposed SPM in Chap. 5 to solve the above model, the optimal trajectory and the corresponding control can be obtained.

## 9.4 Application Case 1: Trajectory Planning Problem Based on the Kinematic Model of UGS

### 9.4.1 Kinematic Model of Unmanned Ground Vehicle

An illustration of an unmanned ground vehicle (UGV) is given in Fig. 9.1. Point $C(x_C, y_C)$ is the mass center of the platform and $\theta$ represents the angle between the

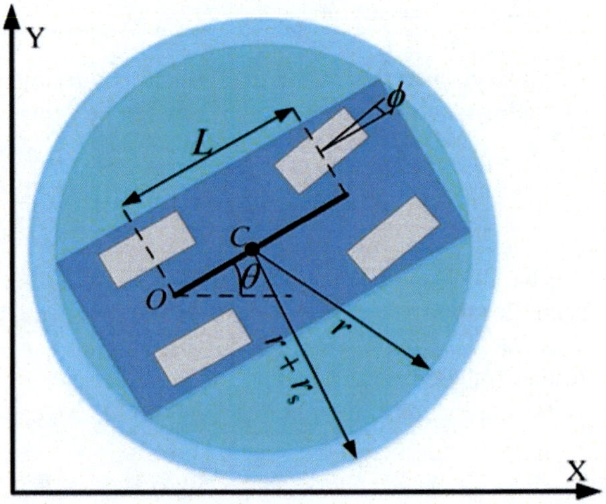


Fig. 9.1 An illustration of UGV

X-axis and the heading direction. Point $O(x, y)$ is the intersection of the wheel axis with the symmetry axis of the platform, and the distance between $C$ and $O$ is $d$. And $x_C = x + d\cos\theta$, $y_C = y + d\sin\theta$.

The sideslip between tires and the ground can be ignored when the speed is low. It assumed that the vehicle can't move sideways, then we have the following nonholonomic constraint

$$\dot{x}(t)\sin\theta - \dot{y}(t)\cos\theta = 0 \tag{9.8}$$

It's a typical nonholonomic system, the constraint of velocity can't be converted to the position. And the position and velocity of the system are constrained, where the corresponding constraints are time-dependent.

For each steering moment, according to Akermann Steering principle, the linear velocity of the vehicle is always tangent to the instantaneous trajectory. Through mathematical deduction, the relationship between the corresponding steering angle input $\alpha$, curvature radius $r$ and the curvature $c$ can be obtained as follows:

$$r = \frac{L}{\tan\alpha} \tag{9.9}$$

$$c = \frac{1}{r} \tag{9.10}$$

Based on the above three equations, the vehicle kinematics model is obtained as follows:

$$\begin{cases} \dot{x} = v\cos\theta \\ \dot{y} = v\sin\theta \\ \dot{\theta} = v\tan\alpha / L \end{cases} \tag{9.11}$$

For many scenarios, we have strict constraints on the velocity of the vehicle at the terminal time, such as autonomous parking for automobiles. Thus, the velocity can be introduced into the state variable, and the following vehicle kinematics model can be obtained

$$\begin{cases} \dot{x} = v\cos\theta \\ \dot{y} = v\sin\theta \\ \dot{\theta} = vu_1 / L \\ \dot{v} = u_2 \end{cases} \tag{9.12}$$

where $\mathbf{X} = (x, y, \theta, v)^{\mathrm{T}}$ is the state variable list, $\mathbf{U} = (u_1, u_2)^{\mathrm{T}} = (\tan\alpha, a)^{\mathrm{T}}$ is the control variable list.

### 9.4.2 Mechanical Constraints

(1) Control constraints

The control input has a great impact on the system, it's necessary to consider the range of control in trajectory planning problem according to the mechanical structure and safety requirements of the system. Specifically, the acceleration of the system and the steering angle of the front wheel should be considered, leading to the following constraints

$$\begin{cases} \tan\alpha_{\min} \le u_1 \le \tan\alpha_{\max} \\ u_{2\min} \le u_2 \le u_{2\max} \end{cases} \tag{9.13}$$

(2) Velocity constraints

From a safety perspective, the velocity of the system should be restrained within a safe range in case of emergency brake,

$$v_{\min} \le v \le v_{\max} \tag{9.14}$$

where the variable $v_{\max}$ and $v_{\min}$ is the maximum and minimum allowable velocity, respectively. And the mathematical description of the avoidance constraint is Eq. (9.5).

### 9.4.3 Numerical Simulations

Now we consider the path planning of a UGV described in Eq. (9.12), and the parameters are set as given in Table 9.1.

And the "time-energy optimal" objective function will be used in this experiment, where the terminal time is unspecific and need to be optimized. Then the SPM for

**Table 9.1** Parameters list of model

| Parameter | Value | Unit |
|---|---|---|
| $L$ | 7 | m |
| $d$ | 0 | m |
| $v_{\max}$ | 1 | m/s |
| $v_{\min}$ | 0 | m/s |
| $\alpha_{\max}$ | 45 | deg |
| $\alpha_{\min}$ | -45 | deg |
| $u_{2\max}$ | 1 | m/s$^2$ |
| $u_{2\min}$ | $-1$ | m/s$^2$ |

**Table 9.2** Parameters list of obstacles

| No. | The $x$ coordinate of the center of the obstacle | The $y$ coordinate of the center of the obstacle |
|---|---|---|
| 1 | 20 | 20 |
| 2 | 70 | 20 |
| 3 | 120 | 20 |
| 4 | 20 | 70 |
| 5 | 70 | 70 |
| 6 | 120 | 70 |
| 7 | 20 | 120 |
| 8 | 70 | 120 |
| 9 | 120 | 120 |

problems with unspecific terminal time developed in Sect. 5.5, is used for numerical solving. To balance the importance of terminal time and energy term, the value of $k$ in Eq. (9.1), can't be too large. In this case, we set $k = 0.05$ (the value is determined by the different understanding of the importance for time and energy index within the right range).

The environment with nine static obstacles are designed in this experiment, the details are shown as Table 9.2. The radius of each obstacle is 10 m, the radius of the system is 9 m, the safe distance is 1 m, and $t_0 = 0$ s. And the corresponding initial boundary conditions and terminal boundary conditions are

$$\mathbf{X}_0 = \left(0,\ 0,\ 90^\circ,\ 0\right)^{\mathrm{T}} \tag{9.15}$$

$$\mathbf{X}_f = \left(90,\ 90,\ 0^\circ,\ 0\right)^{\mathrm{T}} \tag{9.16}$$

The parameters of SPM in this example are set as follows, the whole-time domain is divided into 10 intervals, and a 9-degree approximation is applied to each interval. The convergent tolerance is set as $\varepsilon = 10^{-6}$. As for the iterative method to determine the terminal time, corresponding parameters are set as $\varepsilon' = 10^{-4}$ and $\tau = 10^{-3}$

The final solution can be obtained by 9 iterations using 3.5750 s, the calculated performance index is $J^* = 9.4378$, the terminal time is 148.7184 s.

The control variables in Fig. 9.2 are relatively stable, and the control variables fluctuate within a very small range. To further verify the optimized results, $v$ is analyzed as follows:

Comparative analysis of Figs. 9.2 and 9.3 suggests that the change of $v$ is smooth. It accelerates from 0 m/s to the maximum speed very quickly, then maintains constant for a long period, and finally reduces to 0 m/s at the terminal phase. In additionally, the constraints on state and control variables are strictly satisfied.

In addition, the optimal trajectory of the system is plotted in Fig. 9.4, the solid black line represents the trajectory of the center, and the red outline represents the envelope of the wheeled robot. It can be concluded that the trajectory obtained is

**Fig. 9.2** An illustration of the control

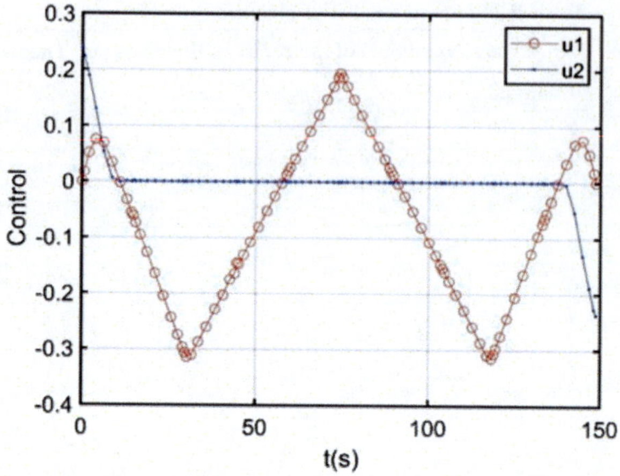


**Fig. 9.3** An illustration of $v$

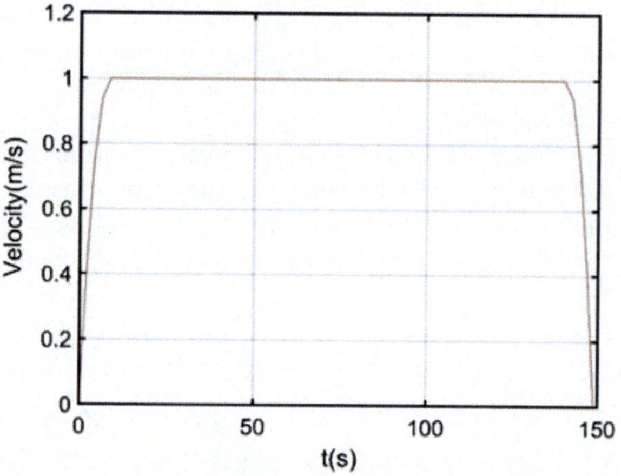


**Fig. 9.4** An illustration of optimal trajectory

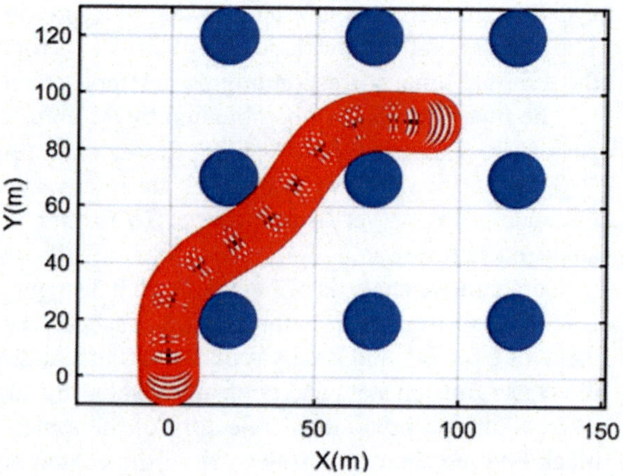

very smooth, and it does not collide with the obstacles in the environment. It shows that the SPM can be effectively applied to the trajectory planning of UGS based on kinematics.

## 9.5 Application Case 2: Trajectory Planning Problem Based on the Dynamic Model of TWMRs

### 9.5.1 *Dynamic Model of TWMRs*

A two-wheel mobile robot (TWMR) is a typical nonholonomic underactuated system, and an illustration of the system is given in Fig. 9.5. Two wheels are driven independently. The width, mass and moment inertia of the platform are denoted as $2b$, $m_b$, and $I_b$, respectively. And the radius, mass and the moment inertia of each wheel are denoted as $r$, $m_w$, and $I_w$, respectively. Let $\varphi_l$ and $\varphi_r$ represent the angular displacements of the left and the right wheels. Let the control vector $\mathbf{U} = [T_r\ T_l]^{\mathrm{T}}$ represent the inserted torque of the two actuators. If one selects the generalized coordinates of the system as

$$\mathbf{x} = [x_C\ y_C\ \theta\ \varphi_r\ \varphi_l\ \dot{\varphi}_r\ \dot{\varphi}_l]^{\mathrm{T}} \tag{9.17}$$

Then the system dynamics can be expressed as

$$\dot{\mathbf{x}} = \mathbf{\Lambda}(\mathbf{x}) + \mathbf{\Gamma}(\mathbf{x})\mathbf{U} \tag{9.18}$$

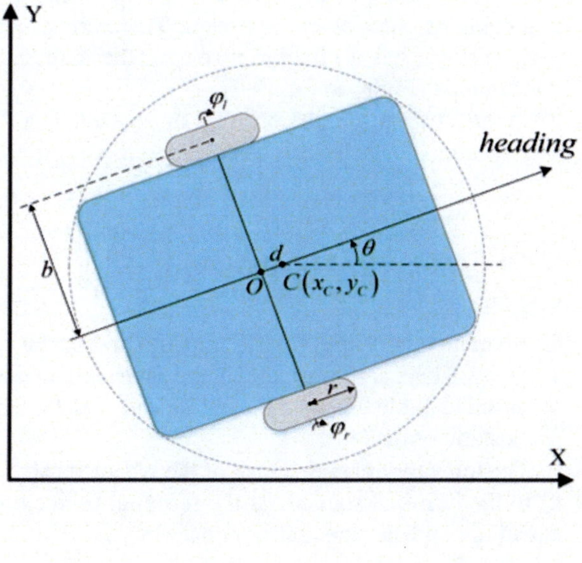


**Fig. 9.5** An illustration of the TWMR

where the detailed expressions of $\mathbf{\Lambda}(\mathbf{x})$ and $\mathbf{\Gamma}(\mathbf{x})$ can be seen in Ref. [18].

### 9.5.2 *Inequality Constraints*

In addition to satisfying the dynamic equation, it also needs to satisfy several constraints. Specifically, consider the following types of constraints in this example,

(1) Control constraints

To avoid actuator saturation, the following constraints should be satisfied

$$|T_l| \leq T_{\max},\ |T_r| \leq T_{\max} \tag{9.19}$$

where the variable $T_{\max}$ is the upper of the input torque of the actuator.

(2) Velocity constraints

If the velocity of the wheeled robot is too fast, it cannot achieve rapid braking when it encounters emergency braking. From a safety perspective, the following constraint is considered

$$\dot{x}^2 + \dot{y}^2 - v_{\max}^2 \leq 0 \tag{9.20}$$

where the variable $v_{\max}$ is the upper of the speed of the wheeled robot.

(3) Avoidance constraints

As shown in Fig. 9.6, it's necessary to consider collision avoidance when it works in a complex working environment. There are $Z$ obstacles in the environment, and each obstacle has a circular envelope, the mathematical description of avoidance constraint is similar to Eq. (9.5).

In this description, the position of obstacle is a function of time, and both static and dynamic obstacles can be considered.

### 9.5.3 *Numerical Simulations*

The parameters of the TWMR model used in the numerical simulation are list in Table 9.3. There are 3 obstacles in the environment, and the corresponding parameters are listed in Table 9.4. And the obstacles 1 and 2 are dynamic obstacles, and obstacle 3 is a static obstacle.

The trajectory planning task of the wheeled robot is to move from the initial state $\mathbf{x}_s$ to the terminal state $\mathbf{x}_f$ at the terminal time $t_f$. At the same time, the dynamic equations and the inequality constraints must be satisfied, and "energy optimal"

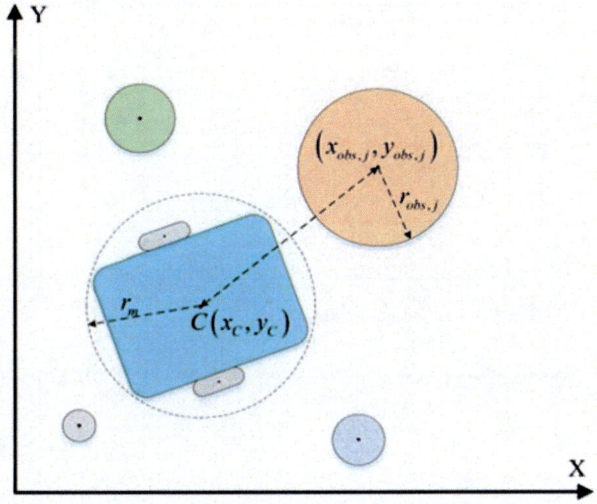


**Fig. 9.6** An illustration of the obstacle avoidance

**Table 9.3** The model parameters used in numerical simulation

| Parameter | Value | Unit |
|---|---|---|
| $d$ | 0 | m |
| $b$ | 0.18 | m |
| $r$ | 0.08 | m |
| $r_m$ | 0.25 | m |
| $r_s$ | 0 | m |
| $m_b$ | 101 | kg |
| $I_b$ | 7.6 | kg·m$^2$ |
| $m_w$ | 6 | kg |
| $I_w$ | 0.04 | kg·m$^2$ |
| $T_{max}$ | 50 | N·m |
| $v_{max}$ | 2 | m/s |

**Table 9.4** Parameters of the Three Obstacles (unit: m)

| No. | Radius | Center |
|---|---|---|
| 1 | 0.05 | $(0.5, 0.1\sin t)$ |
| 2 | 0.05 | $(0.9 + 0.1\sin t, 0.5)$ |
| 3 | 0.05 | $(0.3, 0.7)$ |

objective function will be adopted. The initial time is 0 s and the terminal time is 2 s, and the corresponding initial and terminal boundary conditions are

$$\mathbf{x}_s = (0, 0, 0, 0, 0, 0, 0)^{\mathrm{T}} \tag{9.21}$$

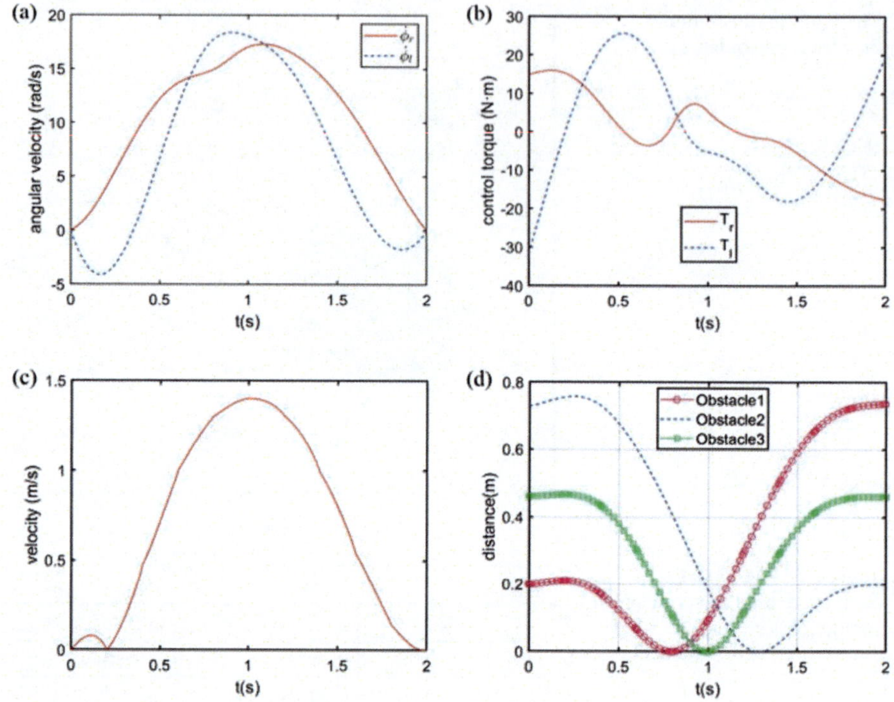


**Fig. 9.7** **a** The angular velocities of wheels, **b** The optimal control inputs, **c** The velocity of the robot, **d** Distances between the robot and obstacles

$$\mathbf{x}_f = \left(1, 1, \pi/2, \text{free}, \text{free}, 0, 0\right)^{\mathrm{T}} \tag{9.22}$$

The algorithm parameters in the SPM are set as $Q = 10$ and $N^{(j)} = 10(j = 1, 2, \cdots, Q)$, and the convergent criterion is $\varepsilon = 1e - 6$. The final solution can be obtained by 13 iterations, calculation time is 5.434 s, and the performance index is $J^* = 1.7703$.

Figure 9.7a denotes the angular velocity of the wheel. And the control input, speed, and the distance between the wheeled robot and obstacle are shown in Fig. 9.7b, Fig. 9.7c, and Fig. 9.7d, respectively. The results show that the constraints are strictly satisfied. In addition, the optimal trajectory is plotted in Fig. 9.8, the solid black line represents the trajectory of the center, and the red outline represents the envelope of the wheeled robot.

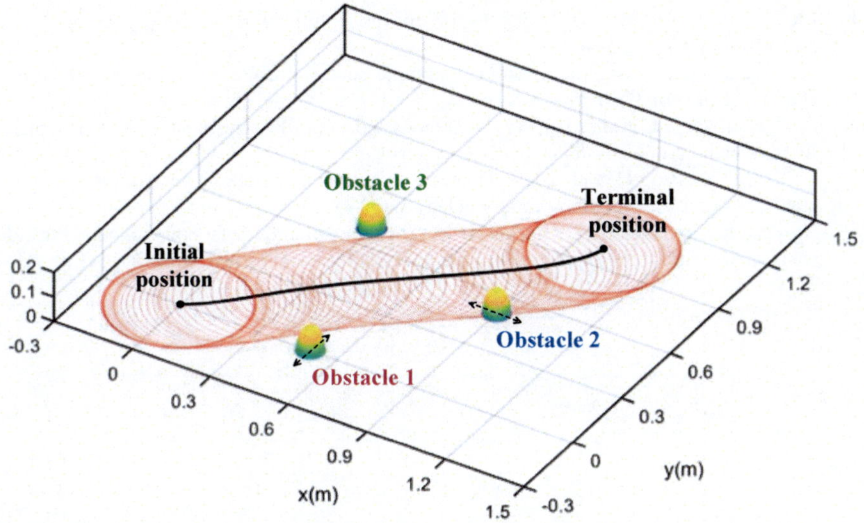


**Fig. 9.8** The optimal trajectory of the TWMR

# References


1. Tong H (2012) Path planning of UAV based on Voronoi diagram and DPSO. Procedia Eng 29:4198–4203
2. Candeloro M, Lekkas A M, Hegde J et al (2016) A 3D dynamic Voronoi diagram-based path-planning system for UUVs. OCEANS 2016 MTS/IEEE Monterey 2016:1–8
3. Wu Y, Qu XJ (2013) Path planning for taxi of carrier aircraft launching. Sci Chine Technol Sci 56(6):1561–1570
4. Wu Y, Qu X (2015) Obstacle avoidance and path planning for carrier aircraft launching. Chin J Aeronaut 28(3):695–703
5. Wu Y, Hu N, Qu X (2019) A general trajectory optimization method for aircraft taxiing on flight deck of carrier. Proc Inst Mech Eng Part G J Aerosp Eng 233(4):1340–1353
6. Zhang Y, Wu L, Wang S (2013) UCAV path planning by fitness-scaling adaptive chaotic particle swarm optimization. Math Prob Eng 2013(2013):705238
7. Das PK, Behera HS, Jena PK et al (2016) Multi-robot path planning in a dynamic environment using improved gravitational search algorithm. J Electr Syst Inf Technol 3(2):1–13
8. Groh K, Röck S (2010) A contribution to collision-free trajectory planning for handling systems in varying environments. Prod Eng Res Dev 4(1):101–106
9. Huptych M, Groh K, Röck S (2011) Online path planning for industrial robots in varying environments using the curve shortening flow method. Intell Robot Appl 2011:73–82
10. Huptych M, Röck S (2015) Online path planning in dynamic environments using the curve shortening flow method. Prod Eng Res Dev 9(5–6):613–621
11. Mancini M, Costante G, Valigi P et al (2017) Towards domain independence for learning-based monocular depth estimation. IEEE Robot Autom Lett 2017
12. Morales N, Toledo J, Acosta L (2016) Path planning using a multiclass support vector machine. Appl Soft Comput 43:498–509
13. Açikmeşe B, Carson JM, Blackmore L (2013) Lossless convexification of nonconvex control bound and pointing constraints of the soft landing optimal control problem. IEEE Trans Control Syst Technol 21(6):2104–2113

14. Liu X, Lu P, Pan B (2017) Survey of convex optimization for aerospace applications. Astrodynamics 1(1):23–40
15. Liu X, Lu P (2014) Solving nonconvex optimal control problems by convex optimization. J Guid Control Dyn 37(3):750–765
16. Liu J, Han W, Liu C et al (2018) A new method for the optimal control problem of path planning for unmanned ground systems. IEEE Access 6:1–1
17. Liu J, Han W, Peng H et al (2019) Trajectory planning and tracking control for towed carrier aircraft system. Aerosp Sci Technol 84(Jan):830–838
18. Nazemizadeh M, Rahimi H, Khoiy K (2012) Trajectory planning of mobile robots using indirect solution of optimal control method in generalized point-to-point Task. Front Mech Eng 7(1):23–28

# Chapter 10
# Motion Planning and Control for Overhead Cranes

## 10.1 Introduction

Overhead crane systems play an important role in many engineering sites, such as harbors, factories, construction sites, etc. Traditionally, overhead crane systems are operated by experienced operators. However, due to the increased demand on fast and accurate payload positioning and suppressing of swing angles, as well as safety issues, automation, and control technologies are overwhelmingly applied in controlling of overhead crane systems. Overhead cranes are essentially an underactuated nonlinear system where motions of trolley of the payload are highly coupled, which makes them difficult to control.

The control techniques for overhead cranes can be generally categorized into two groups, i.e., the open-loop technique and the closed-loop technique. The main difference between these two groups of the method lies in whether the feedback is required. Closed-loop control techniques are less sensitive to external disturbances or parameter variation than open-loop techniques since feedback is used. An excellent review of related control techniques can be referred to [1].

Closed-loop control techniques include proportional-integral-derivative (PID) [2], linear-quadratic regulator (LQR) [3], model predictive control (MPC) [4, 5], adaptive control [6], sliding mode control (SMC) [7], etc. And control-loop techniques fan be further categorized into input shaping [8], adaptive input shaping [9], path planning [10, 11], etc. Open-loop techniques are sometimes integrated with close-loop techniques since their incapability to deal with disturbances and perturbations [4, 11, 12]. It is pointed out that the method mentioned above are all model-based methods. When accurate models are not available, intelligent control methods such as neural network-based methods [13] and fuzzy logic methods [14], can be applied.

The trajectory optimization problem of 3D overhead crane is considered in this chapter, where an "offline planning + online tracking" autonomous motion control



X. Wang et al., *Symplectic Pseudospectral Methods for Optimal Control*, Intelligent Systems, Control and Automation: Science and Engineering 97,
https://doi.org/10.1007/978-981-15-3438-6_10

framework is applied. In the offline planning module, constraints such as input saturation, maximum velocity, maximum sway angle are considered, and the aim is to transfer the system from the initial state to the final state within a fixed time interval. It can be solved under the framework of constrained nonlinear optimal control problems. In the online tracking module, the symplectic pseudospectral MPC is used to track the reference trajectory.

## 10.2 Problem Formulation

### *10.2.1 System Dynamics*

As illustrated in Fig. 10.1, an overhead crane is mainly composed of 4 parts, i.e., trolley, girder, rope and payload. In Fig. 10.1, $x$ and $y$ represent the displacements of the trolley and the girder along the X and Y axes, respectively; $\theta_x$ and $\theta_y$ describe the position of the payload; the rope is considered to be massless and with the length of $l$; $m_g$, $m_t$, and $m_p$ are the masses of the girder, the trolley, and the payload, respectively; and $u_x$ and $u_y$ are the control forces applied to the trolley and the girder, respectively. When one selects the generalized coordination as $\mathbf{q} = \left[x, y, \theta_x, \theta_y\right]^{\mathrm{T}}$, then the dynamic equations of the 3D overhead crane can be derived using the Lagrange's method as follows [15]:

$$\mathcal{M}(\mathbf{q})\ddot{\mathbf{q}} + \mathcal{V}(\mathbf{q}, \dot{\mathbf{q}})\dot{\mathbf{q}} + \mathcal{G}(\mathbf{q}) = \mathcal{F} \tag{10.1}$$

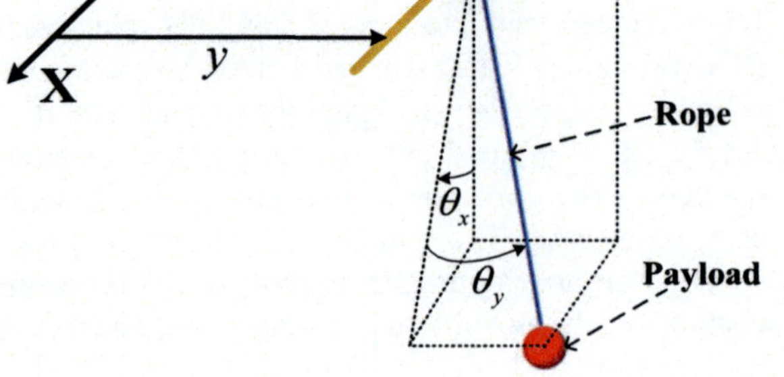


**Fig. 10.1** An illustration of 3D overhead cranes

where $\mathcal{M}(\mathbf{q}) \in \mathbb{R}^{4\times4}$, $\mathcal{V}(\mathbf{q}, \dot{\mathbf{q}}) \in \mathbb{R}^{4\times4}$, $\mathcal{G}(\mathbf{q}) \in \mathbb{R}^4$ are the inertia and the centripetal-Coriolis matrices, respectively; $\mathcal{G}(\mathbf{q}) \in \mathbb{R}^4$ and $\mathcal{F} \in \mathbb{R}^4$ are the gravity and the control vectors, respectively. Their expressions are given as follows:

$$\mathcal{M}=\begin{bmatrix} m_p+m_t & 0 & m_plC_xC_y & -m_plS_xS_y \\ 0 & m_p+m_t+m_g & 0 & m_plC_y \\ m_plC_xC_y & 0 & m_plC_y^2 & 0 \\ -m_plS_xS_y & m_plC_y & 0 & m_pl^2 \end{bmatrix} \tag{10.2}$$

$$\mathcal{V}=\begin{bmatrix} 0\;0 & -m_pl\left(S_xC_y\dot{\theta}_x+C_xS_y\dot{\theta}_y\right) & -m_pl\left(C_xS_y\dot{\theta}_x+S_xC_y\dot{\theta}_y\right) \\ 0\;0 & 0 & -m_plS_y\dot{\theta}_y \\ 0\;0 & -m_pl^2S_yC_y\dot{\theta}_y & -m_pl^2S_yC_y\dot{\theta}_x \\ 0\;0 & m_pl^2S_yC_y\dot{\theta}_x & 0 \end{bmatrix} \tag{10.3}$$

$$\mathcal{G}=\left[0, 0, m_pglS_xC_y, m_pglC_xS_y\right]^{\mathrm{T}} \tag{10.4}$$

$$\mathcal{F}=\left[u_x, u_y, 0, 0\right]^{\mathrm{T}} \tag{10.5}$$

where $g \approx 9.8$ m/s$^2$ is the gravitational acceleration; $C_x := \cos\theta_x$, $S_x := \sin\theta_x$, $C_y := \cos\theta_y$, $S_y := \sin\theta_y$. To apply the optimal control method, Eq. (10.1), is transformed into the following first-order ordinary differential equations as follows:

$$\dot{\mathbf{x}}=\begin{bmatrix}\dot{\mathbf{q}}\\ \ddot{\mathbf{q}}\end{bmatrix}=\begin{bmatrix}\dot{\mathbf{q}}\\ \mathcal{M}^{-1}(\mathcal{F}-\mathcal{G}-\mathcal{V}\dot{\mathbf{q}})\end{bmatrix}=\mathbf{f}(\mathbf{x},\mathbf{u},t) \tag{10.6}$$

where $\mathbf{x} = \left[\mathbf{q}^{\mathrm{T}}, \dot{\mathbf{q}}^{\mathrm{T}}\right]^{\mathrm{T}} = \left[x, y, \theta_x, \theta_y, \dot{x}, \dot{y}, \dot{\theta}_x, \dot{\theta}_y\right]^{\mathrm{T}}$ is the state space and $\mathbf{u} = [u_x, u_x]^{\mathrm{T}}$ is the control input.

### 10.2.2 Constraints

To guarantee the security during the operation, velocities of the trolley and the girder, sway angles and sway angular velocities should be restrained within safety ranges. Such constraints can be defined as follows:

$$|\dot{x}| \le v_{x\max},\ |\dot{y}| \le v_{y\max} \tag{10.7}$$

$$|\theta_x| \le \theta_{x\max},\ |\theta_y| \le \theta_{y\max} \tag{10.8}$$

$$|\dot{\theta}_x| \le \dot{\theta}_{x\max},\ |\dot{\theta}_y| \le \dot{\theta}_{y\max} \tag{10.9}$$

where $v_{x\max}$ and $v_{y\max}$ are the velocity upper limits for the trolley and the girder, respectively; $\theta_{x\max}$ and $\theta_{y\max}$ are the upper limits for swing angles; and $\dot{\theta}_{x\max}$ and $\dot{\theta}_{y\max}$ the upper bounds for angular velocities.

Additionally, constraints on control variables are imposed to avoid input saturations

$$|u_x| \le u_{x\max},\ |u_y| \le u_{y\max} \tag{10.10}$$

where $u_{x\max}$ and $u_{y\max}$ are the upper bounds for $u_x$ and $u_y$, respectively.

### 10.2.3 Control Objectives

A general transportation task of an overhead crane is to transfer the system from the initial state $\mathbf{x}_s$ to the desired state $\mathbf{x}_f$ during a certain time period $[t_s, t_f]$, meanwhile satisfying the constraints in Eqs. (10.7)–(10.10). A two-step control framework is adopted in this chapter: (1) First, a reference trajectory is generated by the offline planning module; (2) Then, the system is driven by the online controller to track the reference trajectory, where an MPC is designed as the core solver.

It is noted that, the constraints in Eqs. (10.7)–(10.10), are considered in both the offline module and the online module.

### 10.2.4 Parameter Settings

The parameters in the system dynamics (10.1) are set as follows:

$$m_t = 20\text{kg},\ m_g = 40\text{kg},\ m_p = 15\text{kg},\ l = 1.5\text{m} \tag{10.11}$$

The objective of the transportation task is to transfer the system from the initial state $\mathbf{x}(t_s) = (0, 0, 0, 0, 0, 0, 0, 0)^{\mathrm{T}}$ to the desired final state $\mathbf{x}(t_f) = (8, 6, 0, 0, 0, 0, 0, 0)^{\mathrm{T}}$, with the initial time $t_s = 0$ and the final time $t_f = 10$. And parameters in the constraints in Eqs. (10.7)–(10.10), are given as follows:

$$v_{x\max} = v_{y\max} = 2\,\text{m/s} \tag{10.12}$$

$$\theta_{x\max} = \theta_{y\max} = 4\,\text{deg} \tag{10.13}$$

$$\dot{\theta}_{x\max} = \dot{\theta}_{y\max} = 8\text{deg}/\text{s} \tag{10.14}$$

$$u_{x\max} = u_{y\max} = 20\text{N} \tag{10.15}$$

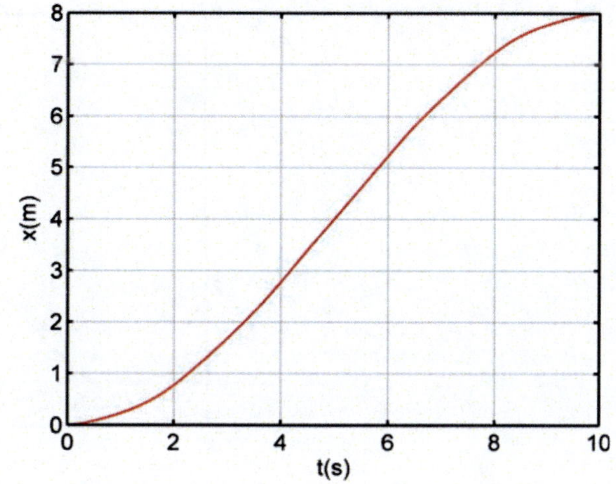


**Fig. 10.2** Profile of $x$

## 10.3 Offline Trajectory Planning

In the offline planning module, the formulate an energy-optimal control problem with the following cost functional

$$J = \int_{t_s}^{t_f} (u_x^2 + u_y^2) \mathrm{d}t \tag{10.16}$$

When using the SPM developed in Chap. 5, to solve this problem, we set the number of sub-intervals as $Q = 20$. And the approximation degree applied in each sub-interval is set as $N^{(j)} = 10$ $(j = 1, 2, \ldots, Q)$. The acceptable tolerance is set to $\varepsilon = 1e - 8$.

Converged solutions can be obtained after 3 iterations and the cost functional is $J = 3.513e3$. Profiles of state and control variables are plotted in Figs. 10.2, 10.3, 10.4, 10.5, 10.6, 10.7, 10.8, 10.9, 10.10 and 10.11, respectively. It is seen that constraints in Eqs. (10.7–(10.10), are well satisfied.

A 3D trajectory of the overhead crane is illustrated in Fig. 10.12.

## 10.4 Online Trajectory Tracking

The open-loop optimal control to be solved at each sampling instant is designed as follows:

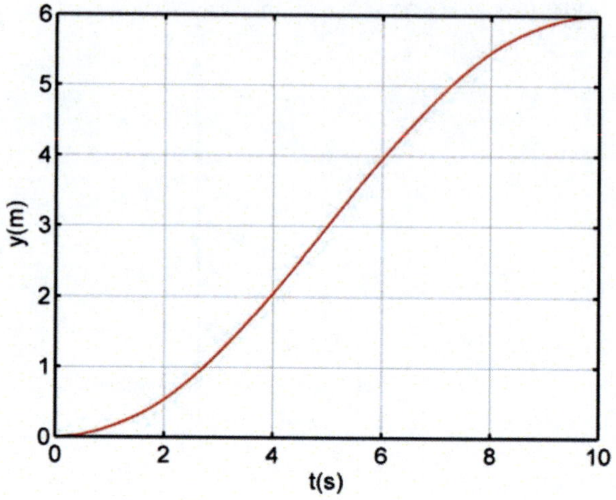


Fig. 10.3 Profile of $y$

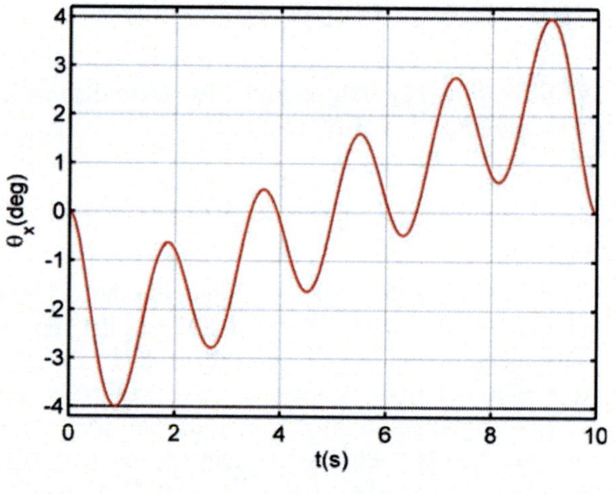


Fig. 10.4 Profile of $\theta_x$

$$\text{Problem } (Q^{[k]}) \begin{cases} \min J = \int_{t_{\Delta k}^0}^{t_{\Delta k}^0 + T} \mathbf{e}^{\mathrm{T}} \tilde{\mathbf{P}} \mathbf{e} + \mathbf{u}^{\mathrm{T}} \tilde{\mathbf{R}} \mathbf{u} dt \\ s.t. \\ \dot{\mathbf{x}} = \tilde{\mathbf{A}} \mathbf{x} + \tilde{\mathbf{B}} \mathbf{u}, \quad k = 1, 2, 3, \ldots \\ \mathbf{x}(t_{\Delta k}^0) = \mathbf{X}_{\Delta k}^0, \ \mathbf{x}(t_{\Delta k}^0 + T) \text{ is free} \\ \mathbf{h}(\mathbf{x}, \mathbf{u}, t) \le \mathbf{0} \end{cases} \tag{10.17}$$

where $\mathbf{e}$ denotes the error between the real trajectory and the reference trajectory; $\tilde{\mathbf{P}}$ and $\tilde{\mathbf{R}}$ weighted matrices on tracking error and control, respectively. For general trajectory tracking problems, no more control inputs are applied after the desired position has arrived. However, for overhead cranes, there may be residual swing angles after the trolley arrives at the desired position. Hence, extra controls are

**Fig. 10.5** Profile of $\theta_y$

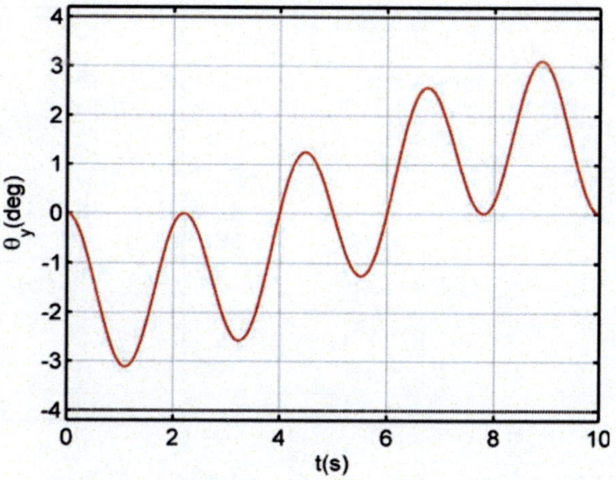


**Fig. 10.6** Profile of $\dot{x}$

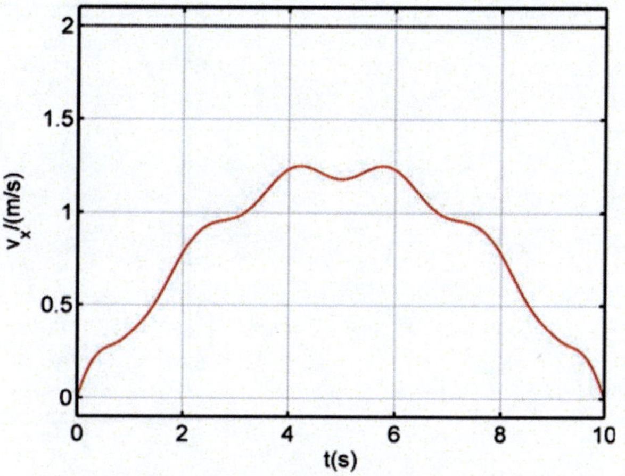


**Fig. 10.7** Profile of $\dot{y}$

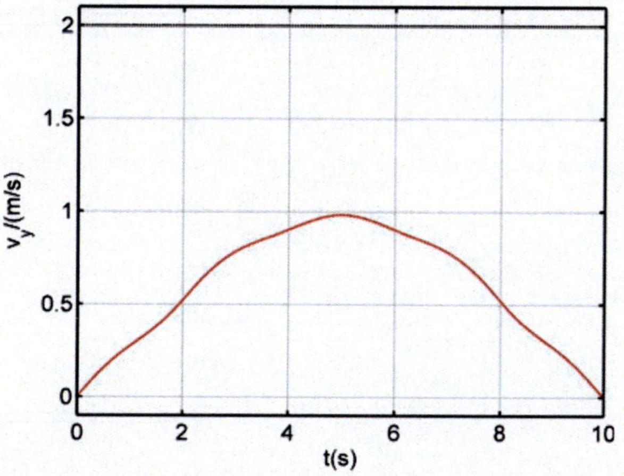

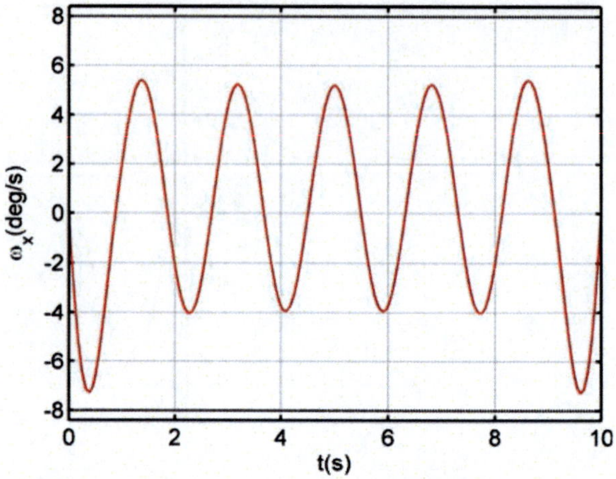


**Fig. 10.8** Profile of $\dot{\theta}_x$

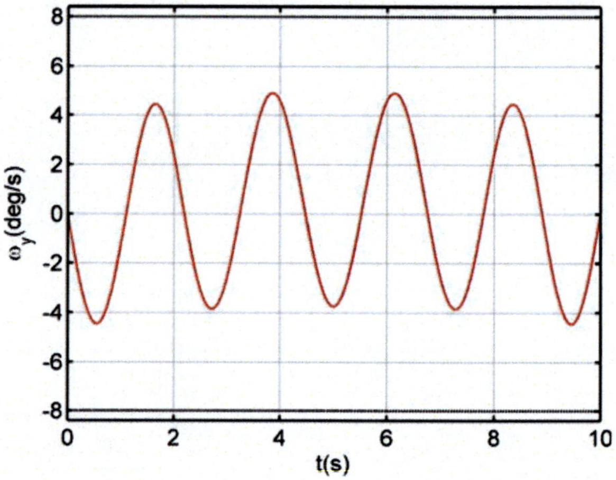


**Fig. 10.9** Profile of $\dot{\theta}_y$

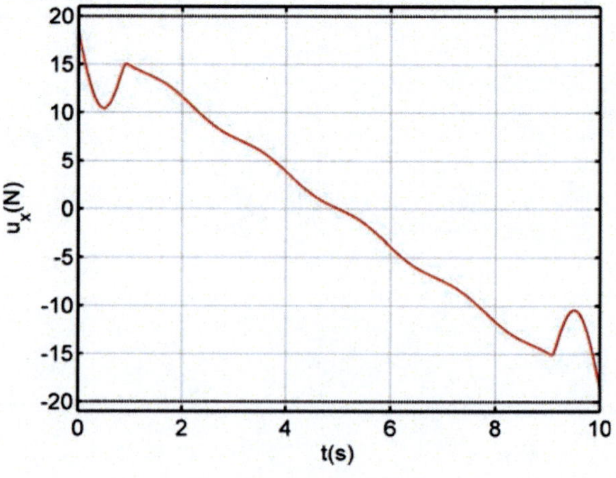


**Fig. 10.10** Profile of $u_x$

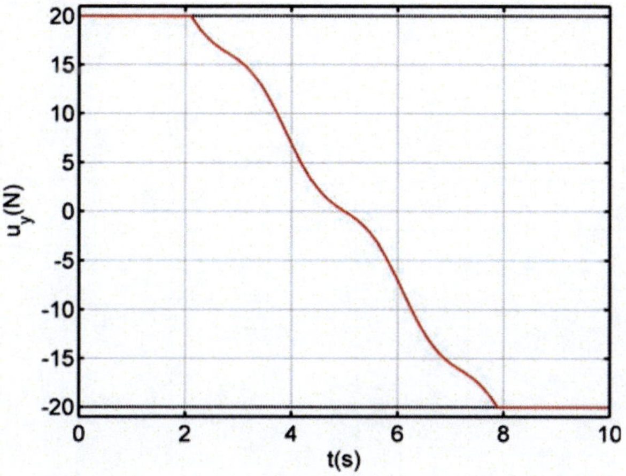


**Fig. 10.11** Profile of $u_y$

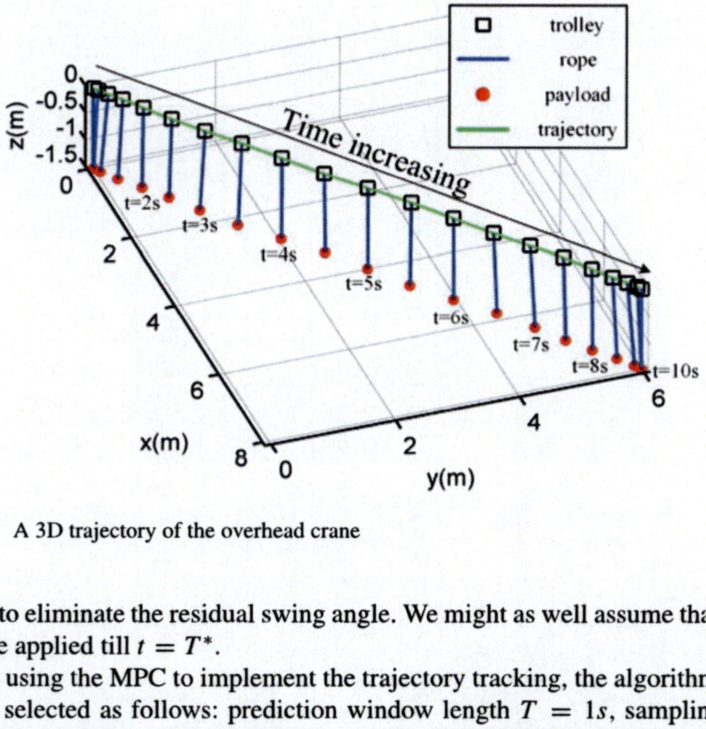


**Fig. 10.12** A 3D trajectory of the overhead crane

required to eliminate the residual swing angle. We might as well assume that control inputs are applied till $t = T^*$.

When using the MPC to implement the trajectory tracking, the algorithm parameters are selected as follows: prediction window length $T = 1s$, sampling period $\delta T = 0.02s$ and final control instant $T^* = 30s$. The SPM developed in Chap. 5 is taken as the core solver. In the SPM, we use only one sub-interval with a 6-degree approximation polynomial. Moreover, the constraints are the same as those in the trajectory planning module. To evaluate the tracking performance, we define the

**Table 10.1** Control performances corresponding to $\tilde{\mathbf{P}}_i (i = 1, 2, 3, 4)$

| No. | Parameter value | $\max_{t \geq 10} \|\mathbf{e}\|_2 (mm)$ | $\max_{t>10} \|\theta_x\| (deg)$ | $\max_{t>10} \|\theta_y\| (deg)$ |
|---|---|---|---|---|
| 1 | $\tilde{\mathbf{P}}_1 \equiv \text{diag}(1e7, 1e7, 1e3, 1e3, 1e1, 1e1, 1e1, 1e1)$ | 5.275 | 0.2731 | 0.7795 |
| 2 | $\tilde{\mathbf{P}}_2(t) = \begin{cases} \text{diag}(1e7, 1e7, 1e3, 1e3, 1e1, 1e1, 1e1, 1e1), t < 10 \\ \text{diag}(1e7, 1e7, 1e5, 1e5, 1e1, 1e1, 1e1, 1e1), t \geq 10 \end{cases}$ | 5.311 | 0.2672 | 0.7430 |
| 3 | $\tilde{\mathbf{P}}_3(t) = \begin{cases} \text{diag}(1e7, 1e7, 1e3, 1e3, 1e1, 1e1, 1e1, 1e1), t < 10 \\ \text{diag}(1e7, 1e7, 1e6, 1e6, 1e1, 1e1, 1e1, 1e1), t \geq 10 \end{cases}$ | 5.580 | 0.2314 | 0.6820 |
| 4 | $\tilde{\mathbf{P}}_4(t) = \begin{cases} \text{diag}(1e7, 1e7, 1e3, 1e3, 1e1, 1e1, 1e1, 1e1), t < 10 \\ \text{diag}(1e7, 1e7, 1e7, 1e7, 1e1, 1e1, 1e1, 1e1), t \geq 10 \end{cases}$ | 14.88 | 0.1405 | 0.6061 |

position error vector, i.e.,$\mathbf{e} = (x_{\text{real}} - x_{\text{ref}}, y_{\text{real}} - y_{\text{ref}})$. And $\|\mathbf{e}\|_2$ shows the distance between the real and the reference trajectories.

### *10.4.1 Selection of Weighted Matrices*

In this sub-section, the initial conditions are identical to those in offline trajectory planning. To effectively eliminate the residual swing angles, we adopt a piece-wise weighted factors on swing angles. The weighted matrix on control is fixed as $\tilde{\mathbf{R}} = diag(1, 1)$; and four different weighted matrices on error can be selected as shown in Table 10.1. For $\tilde{\mathbf{P}}_1$, the weighted matrix keeps constant during the whole control horizon. However, for $\tilde{\mathbf{P}}_i (i = 2, 3, 4)$, the weighted factors on swing angles are increased after $t = 10s$. The profiles of tracking error and swing angles using different $\tilde{\mathbf{P}}_i (i = 1, 2, 3, 4)$ are plot in Figs. 10.13, 10.14 and 10.15 and summarized in Table 10.1. In Figs. 10.13, 10.14 and 10.15, results for $\tilde{\mathbf{P}}_i (i = 1, 2, 3, 4)$ are represents by the red solid line, the green dash line, the blue dotted line, and the purple dash-dotted line, respectively.

From Table 10.1 and Figs. 10.13, 10.14 and 10.15, we can summarize that

(i) All four values of the weighted matrix on error can result in good positioning and sound swing elimination performances;
(ii) After $t = 10s$, the amplitudes of swing angles decrease as time goes over. Moreover, when increasing the weighted factors on swing angles, the amplitudes of swing angles decrease. Furthermore, the residual swing angles converge to zeros faster if the weighted factors on swing angles are bigger.

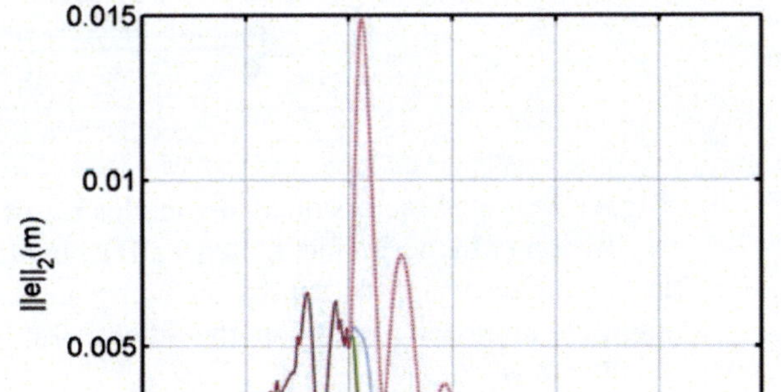


**Fig. 10.13** Profiles of $\|\mathbf{e}\|_2$ under different $\tilde{\mathbf{P}}_i$

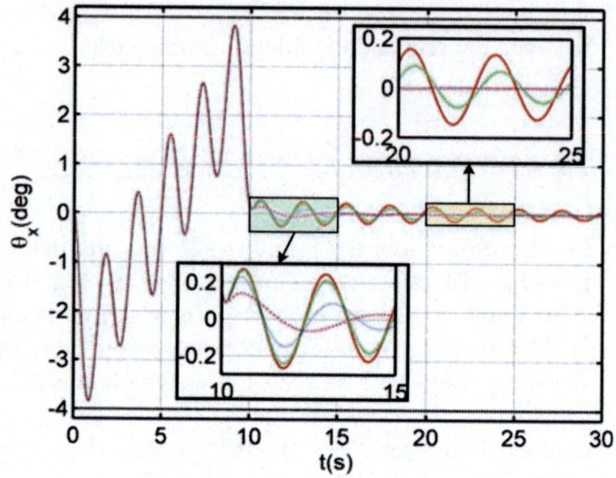

**Fig. 10.14** Profiles of $\theta_x$ under different $\tilde{\mathbf{P}}_i$

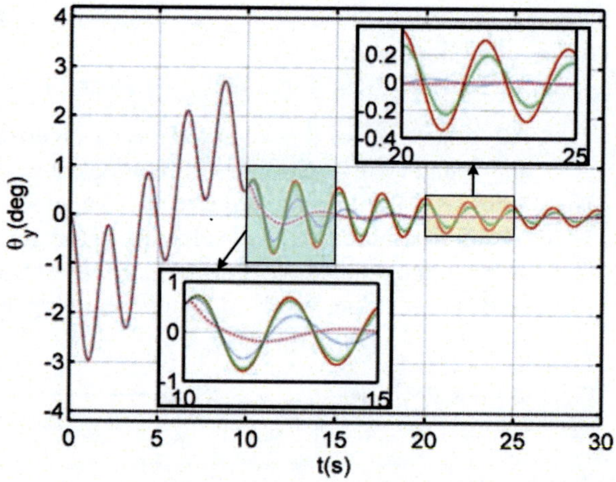

**Fig. 10.15** Profiles of $\theta_y$ under different $\tilde{\mathbf{P}}_i$

(iii) Bigger weighted factors on swing angles result in a relatively bigger position oscillation (which is in fact quite small) right after $t = 10s$; however, as stated in (2), bigger weighted factors on swing angles result in less time for swing angles to approach to zeros, in the other word, it takes less time for the trolley finally stops.

Based on the above simulation, we consider $\tilde{\mathbf{P}}_3(t)$ in Table is a good choice for weighted matrix on error. In the following simulation, unless otherwise stated, the weighted matrices on error and control are select as $\tilde{\mathbf{P}} = \tilde{\mathbf{P}}_3(t)$ and $\tilde{\mathbf{R}} = \text{diag}(1, 1)$, respectively.

**Table 10.2** Perturbations in initial conditions for eight cases

| Case | Initial conditions | | | | | | | |
|---|---|---|---|---|---|---|---|---|
| | $x(m)$ | $y(m)$ | $\theta_x(deg)$ | $\theta_x(deg)$ | $v_x$(m/s) | $v_y$(m/s) | $\omega_x(deg/s)$ | $\omega_y(deg/s)$ |
| 1 | −0.05 | −0.05 | 0 | 0 | 0 | 0 | 0 | 0 |
| 2 | −0.1 | −0.1 | 0 | 0 | 0 | 0 | 0 | 0 |
| 3 | −0.2 | −0.2 | 0 | 0 | 0 | 0 | 0 | 0 |
| 4 | −0.4 | −0.4 | 0 | 0 | 0 | 0 | 0 | 0 |
| 5 | 0 | 0 | −0.5 | −0.5 | 0 | 0 | 0 | 0 |
| 6 | 0 | 0 | −1 | −1 | 0 | 0 | 0 | 0 |
| 7 | 0 | 0 | −2 | −2 | 0 | 0 | 0 | 0 |
| 8 | 0 | 0 | −3 | −3 | 0 | 0 | 0 | 0 |

### *10.4.2 Robust Against Perturbations in Initial Conditions*

In this scenario, we test the control performance for the MPC in the presence of perturbations in initial conditions which are summarized in Table 10.2. It is seen that Cases 1–4, have perturbations in initial position while Cases 5–8, have perturbation in initial swing angles.

The results of Cases 1–4, are plotted in Fig. 10.16, and results for Cases 5–8, are plotted in Fig. 10.17. Simulations on perturbations in initial positions show that the MPC reduces the error between the real and the reference trajectory quickly. Simulations on perturbations in initial swing angles show that the MPC exhibits excellent position tracking performance. It is seen that perturbations in initial position result in much bigger residual swing angles than those for perturbations in initial swing angles; moreover, the bigger the perturbations are, the bigger the residual swing angles are. However, the proposed trajectory tracking controller shows excellent control performance for both tracking position and eliminating residual swing angles.

### *10.4.3 Robust Against Abrupt Variation of the Mass of the Payload*

In this scenario, simulations in the presence of abrupt variation of the mass of the payload are carried out. It is assumed that the $m_p$ varies from 15 kg to 7.5 kg abruptly at $t = 2$ s. Profiles of the state variables are plotted in Figs. 10.18, 10.19, 10.20 and 10.21, and profiles of control inputs are plotted in Fig. 10.22. It is seen that constraints on states and controls are strictly satisfied; the overhead crane tracks the reference trajectory well and the swing angle is well suppressed.

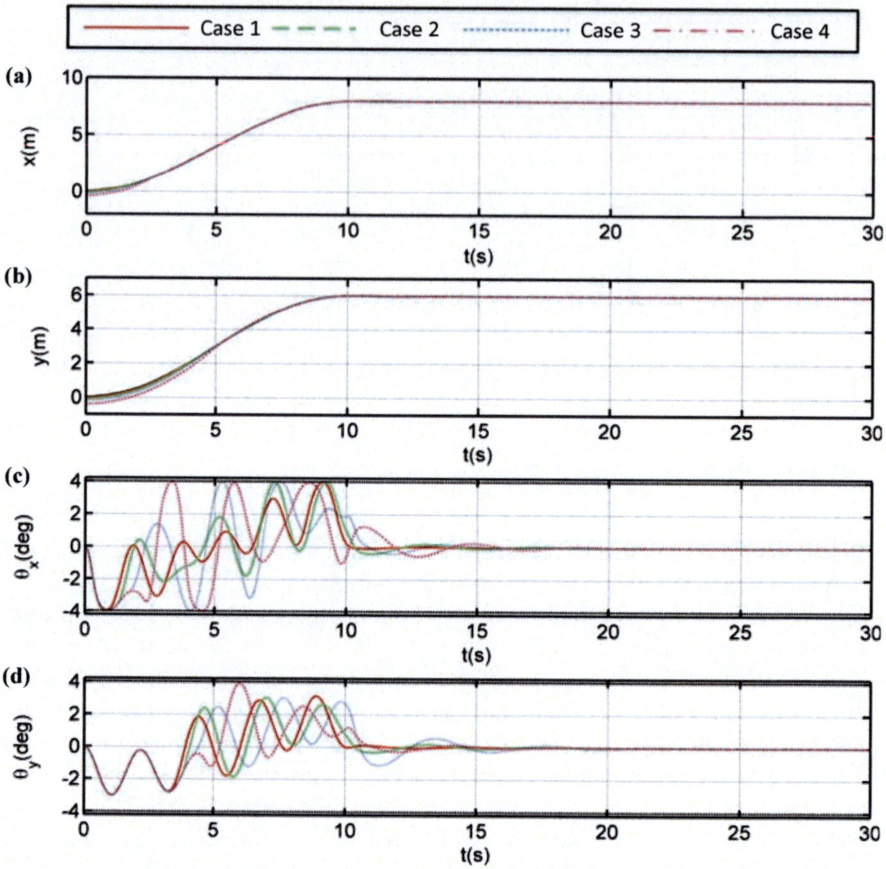


**Fig. 10.16** Profiles of **a** position of the trolley, **b** position of the girder, **c** $\theta_x$ and **d** $\theta_y$ in the presence of perturbations in the initial positions

### *10.4.4 Robust Against Various External Disturbances*

In this scenario, simulations in the presence of various external disturbances on swing angles are carried out. External disturbances are imposed as follows: (1) to simulate the possible collision between of payload and obstacles, impulse disturbances (with an amplitude of 2° to $\theta_x$ and 2° to $\theta_y$) are imposed at $t = 2$s; (2) to simulate the possible influence of wind, continuous random disturbances with amplitudes of 1° are imposed between 15 s and 17s. In addition, the partial feedback linearization (PFL) controller (the control gains are fully tuned and chosen as $k_a = 1$, $k_\varepsilon = 1$, $k_\theta = 0.5$) developed in [15], is taken for comparison.

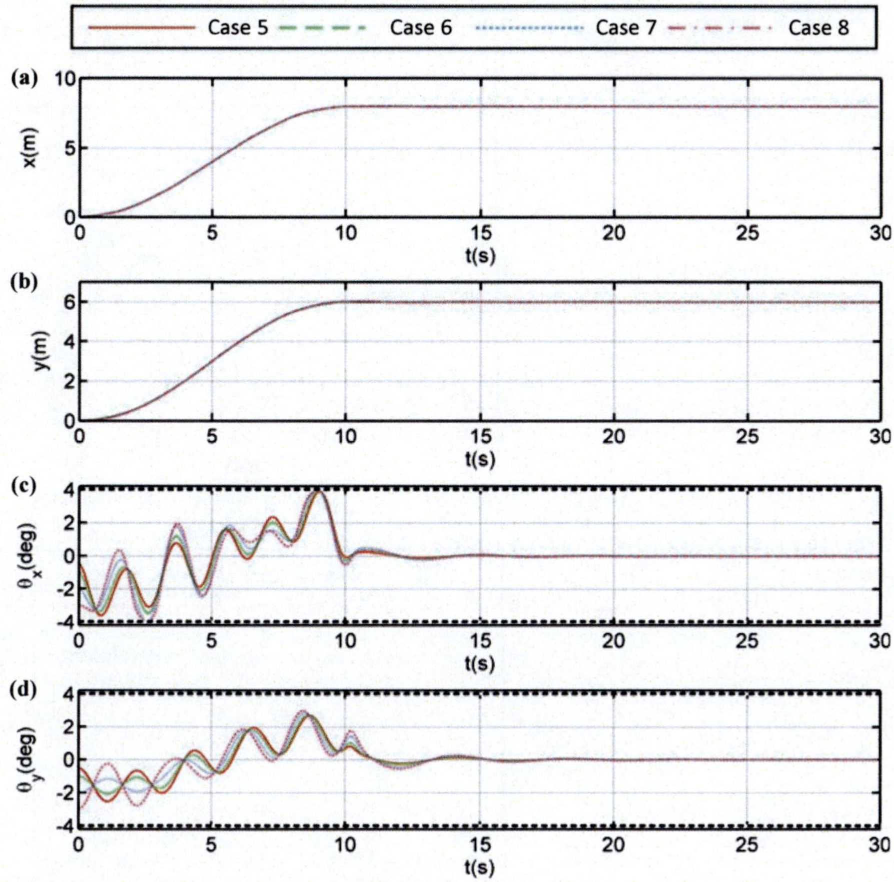


**Fig. 10.17** Profiles of **a** position of the trolley, **b** position of the girder, **c** $\theta_x$, and **d** $\theta_y$ in the presence of perturbations in the initial swing angles

Profiles of state variables and control variables are given in Figs. 10.23 and 10.24, respectively. From Fig. 10.23a, b, it can be seen that the MPC controller tracking the reference trajectory better than the PFL controller. At $t = 10$s, the position obtained by the PFL controller exceeds the desired position significantly and it takes more time to arrive at the desired position. As for the proposed MPC controller, it doesn't make much deviation from the reference trajectory during the whole transportation even when the disturbances work. From Fig. 10.23c, d, it is seen that no matter for 10s $< t <$ 15s or $t >$ 17s, the proposed MPC controller eliminates residual angles more quickly than the PFL controller. From Fig. 10.24, it is seen that after $t = 10$s, control inputs of the MPC controller quickly converge around zero; however, as for

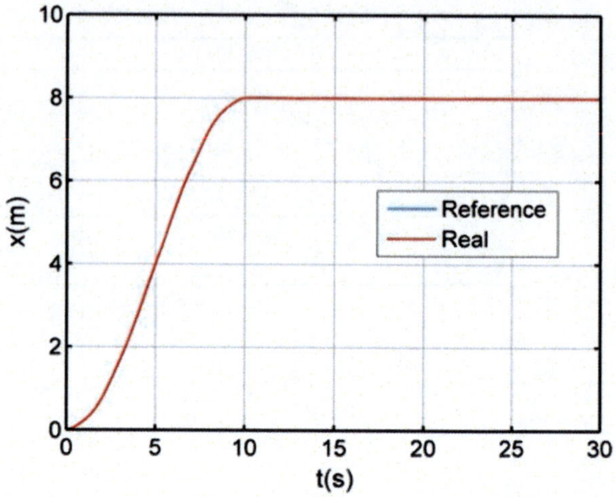


**Fig. 10.18** The profile of state $x$

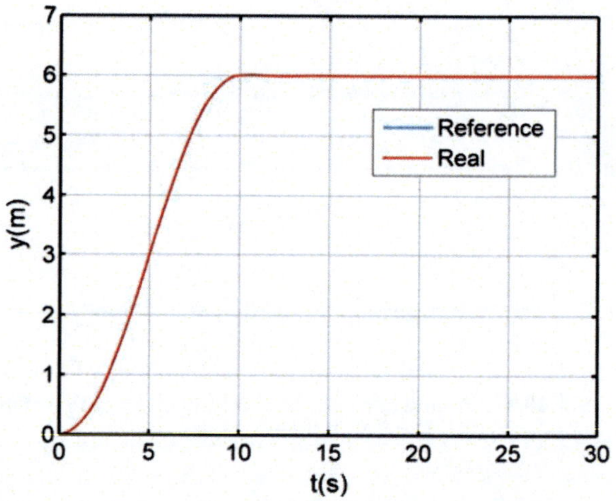


**Fig. 10.19** The profile of state $y$

the PFL controller, there're bigger residual swing angles and a relatively big deviation from the desired position at $t = 10$s, more control inputs are required. Observe the profiles during $15\text{s} < t < 17\text{s}$ and we find that though the control inputs of the MPC controller are bigger than those of the PFL controller, they converge around zero more quickly than those in the PFL controller. Though the control inputs of the MPC controller are relatively big when the continuous disturbances are imposed, they don't exceed the control limits.

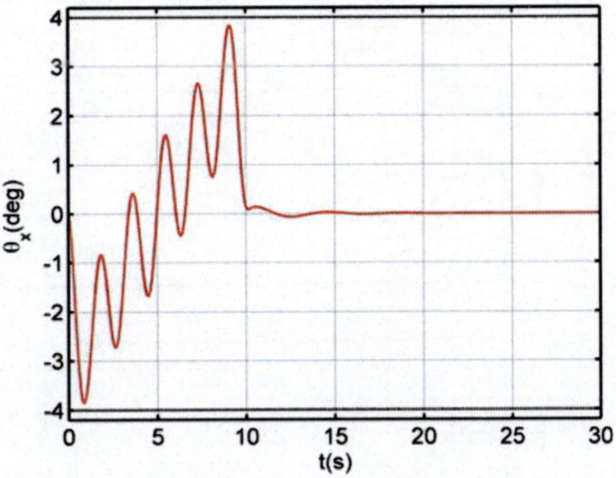


**Fig. 10.20** The profile of state $\theta_x$

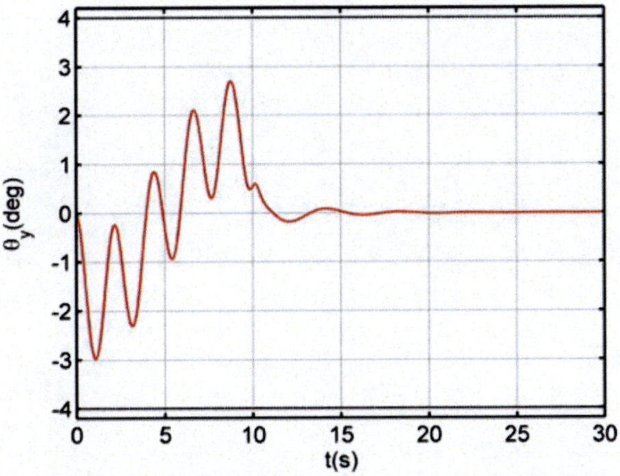


**Fig. 10.21** The profile of state $\theta_y$

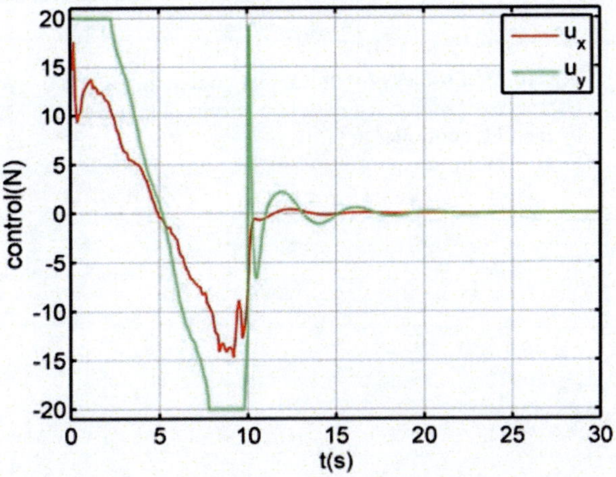


**Fig. 10.22** The profiles of control inputs

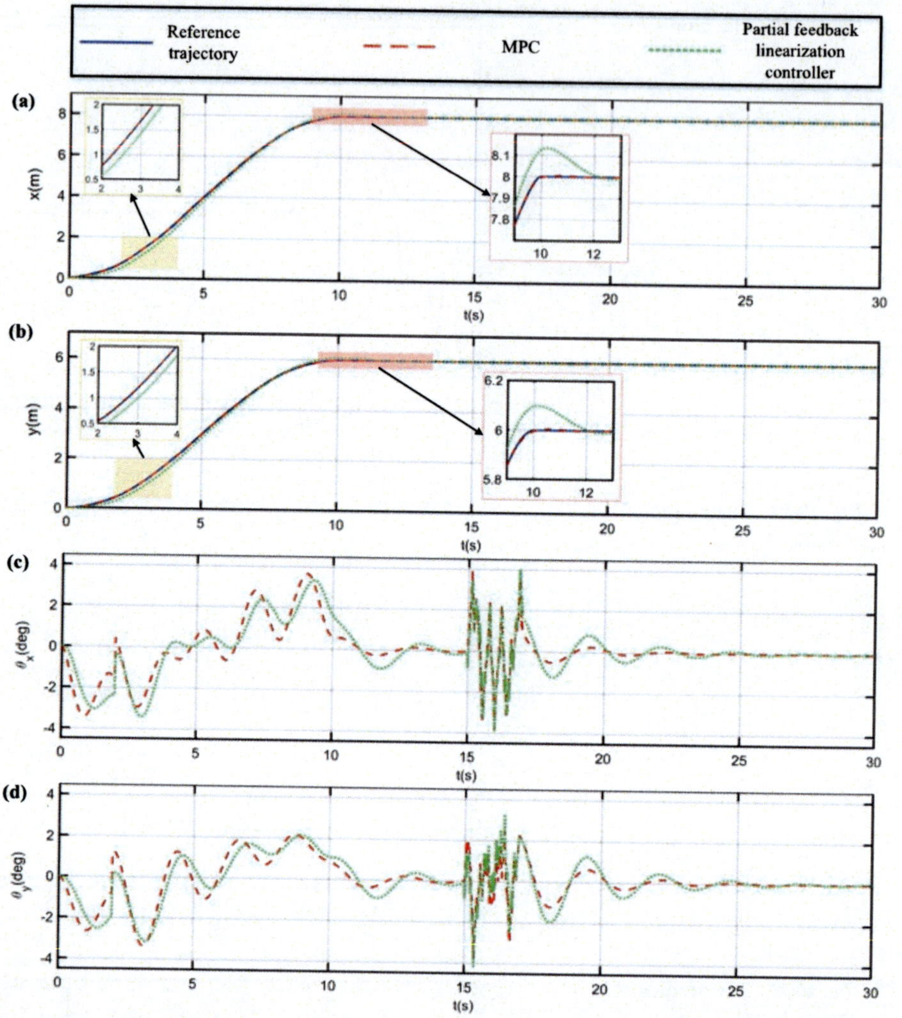


**Fig. 10.23** Profiles of state variables, including **a** $x$, **b** $y$, **c** $\theta_x$, and (d) $\theta_y$. (Blue solid line in (a–b): reference trajectory; red dash line: results obtained by the MPC; green dotted line: results obtained by the PFL controller in [15])

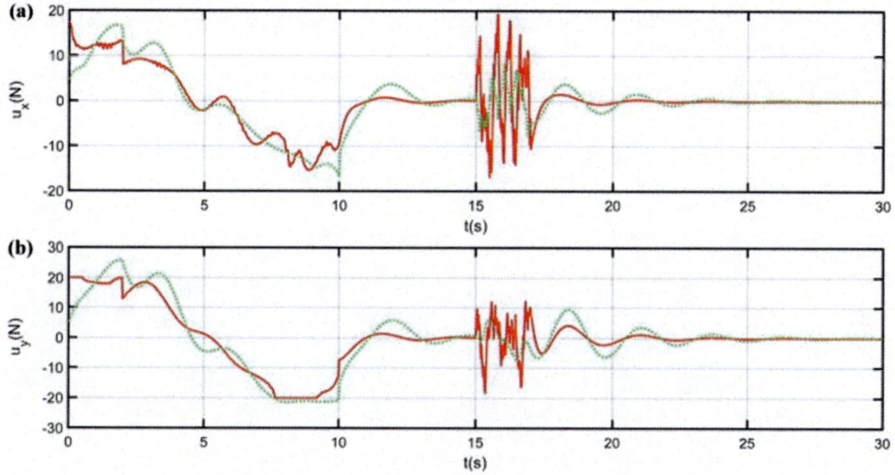


**Fig. 10.24** Profiles of control variables, including **a** $u_x$ and **b** $u_y$. (Red solid line: results obtained by the MPC; green dotted line: results obtained by the PFL controller in [15])

# References


1. Ramli L, Mohamed Z, Abdullahi AM et al (2017) Control strategies for crane systems: a comprehensive review. Mech Syst Signal Process 95:1–23
2. Jaafar HI, Hussien SYS, Ghazali R (2015) Optimal tuning of PID + PD controller by PFS for gantry crane system. The 10th Asian Control Conf. Sabah
3. Yang B, Xiong B (2010) Application of LQR techniques to the anti-sway controller of overhead crane. Adv Mater Res 139–141:1933–1936
4. Chen H, Fang Y, Sun N (2016) A swing constraint guaranteed MPC algorithm for Underactuated overhead cranes. IEEE/ASME Trans Mechatron 21(5):2543–2555
5. Wu Z, Xia X, Zhu B (2015) Model predictive control for improving operational efficiency of overhead cranes. Nonlinear Dyn 79(4):2639–2657
6. Fang Y, Ma B, Wang P et al (2012) A motion planning-based adaptive control method for an underactuated crane system. IEEE Trans Control Syst Technol 20(1):241–248
7. Almutairi NB, Zribi M (2009) Sliding mode control of a three-dimensional overhead crane. J Vib Control 15(11):1679–1730
8. Garrido S, Abderrahim M, Gimenez A et al (2008) Anti-swinging input shaping control of an automatic construction crane. IEEE Trans Autom Sci Eng 5(3):549–557
9. Cole MOT (2011) A discrete-time approach to impulse-based adaptive input shaping for motion control without residual vibration. Automatica 47(11):2504–2510
10. Zhang X, Fang Y, Sun N (2014) Minimum-time trajectory planning for underactuated overhead crane systems with state and control constraints. IEEE Trans Industr Electron 61(12):6915–6925
11. Chen H, Fang Y, Sun N (2016) Optimal trajectory planning and tracking control method for overhead cranes. IET Control Theory Appl 10(6):692–699
12. Wang X, Liu J, Zhang Y et al (2019) A unified symplectic pseudospectral method for motion planning and tracking control of 3D underactuated overhead cranes. Int J Robust Nonlinear Control 29(7):2236–2253
13. Lee LH, Huang PH, Shih YC et al (2014) Parallel neural network combined with sliding mode control in overhead crane control system. J Vib Control 20(5):749–760

14. Smoczek J (2014) Fuzzy crane control with sensorless payload deflection feedback for vibration reduction. Mech Syst Signal Process 46(1):70–81
15. Wu X, He X (2016) Partial feedback linearization control for 3-D underactuated overhead crane systems. ISA Trans 65:361–370

# Chapter 11
# Path Planning for Tractor-Trailer System

## 11.1 Introduction

Tractor-trailer systems, which are typical nonholonomic and unactuated mechanical systems, are now widely used in various fields, such as factory transportation, farming, geographic exploration, etc. A tractor-trailer system consists of a tractor and one or more trailers, where the tractor mainly determines the direction and speed of system movement. Due to the existence of multiple bodies, their path planning and control problem is hard to be solved, especially the control of reverse motion, and the traditional control theory is difficult to be applied. With the increase of the number of trailers, the complexity of controller design and the calculation will increase dramatically, the control process becomes very cumbersome, and the results are difficult to generalize to the system with different trailers.

The tractor-trailer system has the motion capability of a mobile robot or vehicle. There are many scholars who have studied the motion control problems of the tractor-trailer system, such as structural optimization problems, path planning problems, and reverse path tracking problems [1–5]. The configuration control problem of the tractor-trailer system requires that the tractor and trailer can be stabilized in a given configuration, which belongs to the problem of feedback stabilization of the system. Since the system has nonholonomic constraints and does not satisfy the necessary conditions of Brockett's theorem, there is no efficient continuous time-invariant asymptotically stable feedback control method. How to find a discontinuous or time-varying feedback control method to solve the configuration control problem of the nonholonomic systems has become a key issue in this field.

The trailer does not have the ability to move actively, and its speed and direction are mainly determined by the tractor. With the development of computer technology and information technology, autonomous unmanned tractor-trailer systems have received more and more attention and application [6, 7]. The speed and movement direction


X. Wang et al., *Symplectic Pseudospectral Methods for Optimal Control*,
Intelligent Systems, Control and Automation: Science and Engineering 97,
https://doi.org/10.1007/978-981-15-3438-6_11

of the tractor is determined by the automatic controller for the unmanned intelligent tractor-trailer system [8–10]. For most application scenarios, the system mainly controls the position of the trailer, not the position of the tractor. As the structure is more complex, the difficulty of controlling the trailer is much greater than that of a single car without a trailer [11, 12].

## 11.2 The Kinematics Model of the Tractor-Trailer System

The trajectory planning of the tractor-trailer system is so delicate that it should be modeled according to its characteristics, and the effect of trajectory planning depends heavily on the kinematic model or dynamic model used, so it's necessary to establish a dynamic or kinematic model that can fully describe its motion characteristics. Similar to the UGS, the dynamic and kinematic model have their own advantages and disadvantages, which are embodied in the following aspects.

The dynamic model of the tractor-trailer system can be accurate to describe the behavior and characteristics of the system, such as the mechanical properties, external force, the inertia force, etc. Its structure is complex, and the multi-body modeling technology should be adopted to obtain the accurate dynamics model. Generally speaking, the dynamic model is more complex than that of UGS, and it involves many parameters, the work of model parameter identification will bring some challenges to trajectory planning, which also brings inconvenience to the application of dynamics model for tractor-trailer system.

The kinematics model of the tractor-trailer system can accurately describe the motion characteristics of the system under the condition of low speed motion. Compared with the dynamic model, the main advantage of kinematics model is that it's simpler than a dynamic model with less parameters to be identified, and it will cost less computational time in trajectory planning problem based on kinematics model. The disadvantage of the kinematic model is that it does not consider the mechanical properties of the system, and it will not apply to the high-speed motion scene. But if the system is operating at a low speed on the earth's surface, these assumptions do not significantly affect the accuracy of the results.

### *11.2.1 Tractor-Trailer System Without Drawbar*

Based on the assumption, the kinematics and corresponding mechanical constraints are given in the rest of this section. It's noted that when the tractor-trailer system is towed by a tractor, the configuration can be categorized into on-axle or off-axle hitching according to whether the hitch between the tractor and the drawbar/tractor-trailer system lies on the rear axle of the tractor. Actually, the off-axle hitching is the generalized form on-axle hitching. Hence, we will take the off-axle hitching configuration for example when we model the traction system.

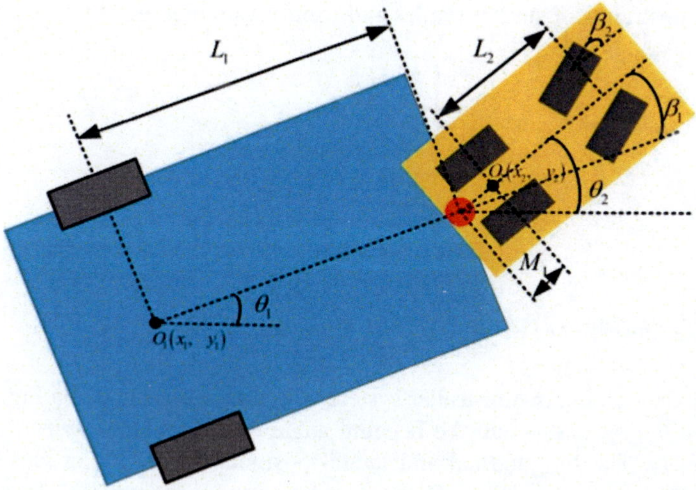


**Fig. 11.1** An example of a typical tractor-trailer system without drawbar

Without considering the lateral thrust, friction or inertia of the system, the kinematic relationship can be simplified, and an example of a typical tractor-trailer system is Fig. 11.1. $\theta_1$ and $\theta_2$ represent the angle between the heading angle of the trailer and tractor, respectively. $O_1(x_1, y_1)$ and $O_2(x_2, y_2)$ represent the center position of the rear wheel of the trailer and tractor, respectively. $v_1$ and $v_2$ represent the translational speed of the rear wheel of the trailer and tractor, respectively. $\beta_1 = \theta_2 - \theta_1$ is the steering angle of the trailer, $\beta_2$ is the steering angle of the tractor, $L_1$ is the length from the front and rear wheel of the trailer, $L_2$ is the length from the front and rear wheel of the tractor, and $M_1$ is the length from the hinge point of the tractor-trailer system and $O_2$. According to the structural characteristics of the traction system, the relationship between $v_1$ and $v_2$, is shown as follows:

$$v_1 = v_2(\cos\beta_1 + L_2 \sin\beta_1 \tan\beta_2 / M_1) \tag{11.1}$$

According to the structural, the kinematic equation can be obtained as follows:

$$\frac{\mathrm{d}}{\mathrm{d}t}\begin{bmatrix} x_1 \\ y_1 \\ \theta_1 \\ \theta_2 \\ v_2 \end{bmatrix} = \begin{bmatrix} (L_2 v_2 \cos\beta_1 + M_1 v_2 u_1 \sin\beta_1)\cos\theta_1 / L_2 \\ (L_2 v_2 \cos\beta_1 + M_1 v_2 u_1 \sin\beta_1)\sin\theta_1 / L_2 \\ (L_2 v_2 \sin\beta_1 - M_1 v_2 u_1 \cos\beta_1) / L_1 L_2 \\ v_2 u_1 / L_2 \\ u_2 \end{bmatrix} \tag{11.2}$$

where the state variables are $\boldsymbol{X} = (x_1, y_1, \theta_1, \theta_2, v_2)^{\mathrm{T}}$, the control variables are $\boldsymbol{U} = (u_1, u_2)^{\mathrm{T}}$,and $u_1 = \tan\beta_2$. The tractor-trailer system is an on-axle traction system when $M_1$ is 0 m; or, it's an off-axle traction system.

The constraints of state variables and control variables are

$$\begin{cases} v_{1\min} \le v_1 \le v_{1\max} \\ |\beta_1| \le \beta_{1\max} \\ |u_1| \le \tan \beta_{2\max} \\ |u_2| \le \alpha_{\max} \end{cases} \quad (11.3)$$

### 11.2.2 Tractor-Trailer System with Drawbar

An off-axle hitching tractor-trailer system with drawbar is shown in Fig. 11.2, where $\theta_1$, $\theta_2$, and $\theta_3$ are the heading angle of the trailer, drawbar, and tractor, respectively. The longitudinal and lateral positions of the trailer can be denoted as $O_1(x_1, y_1)$, the positions of the drawbar can be denoted as $O_2(x_2, y_2)$, the positions of the hinge point of tractor and tractor can be denoted as $O_3(x_3, y_3)$, and the positions of the hinge point of tractor and tractor can be denoted as $O_4(x_4, y_4)$. $\beta_1$, $\beta_2$, and $\alpha$ are the corresponding steering angle, $L_1$ denotes the length from hitch to the rear tire of trailer, $L_2$ denotes the length of drawbar, $M$ denotes length from hitch to the rear tire of tractor, $L_3$ denotes the length from front tire to rear tire of tractor, the velocity of trailer and tractor is $v_1$ and $v_3$, respectively. According to the mechanical design and provisions of equipment, there are some constraints, such as the velocity constraints of trailer and tractor, the constraints of the steering angle. Furthermore, the obstacle avoidance should be taken into consideration.

According to Fig. 11.2, $\beta_1 = \theta_2 - \theta_1$, $\beta_2 = \theta_3 - \theta_2$, and the velocity of trailer $v_1$ and tractor $v_3$ should meet the relationship as follows:

$$v_1 = \frac{v_3 \cos \beta_1 \cos \beta_2 (L_3 + M \tan \beta_2 \tan \alpha)}{L_3} \quad (11.4)$$

Since the goal is to transfer the trailer from the starting position to destination, the trailer is the focus in this chapter. The posture of the trailer should be seen as the state variables so that it's easy to meet the constraints of initial and terminal position. And only another two variables ($\beta_1$,$\beta_2$) are needed, the system can be determined completely. So the state variable is $\mathbf{X} = \begin{pmatrix} x_1 & y_1 & \theta_1 & \beta_1 & \beta_2 \end{pmatrix}^{\mathrm{T}}$, and the control variable can be denoted as $\mathbf{U} = \begin{pmatrix} u_1 & u_2 \end{pmatrix}^{\mathrm{T}}$. The control variable can be selected as the velocity of trailer $u_2$ and the steering angle of tractor $u_1$. To make the model more concise $u_1$ will be replaced by $\tan \alpha$. And the kinematic model of the tractor-trailer system with drawbar is established as follows:

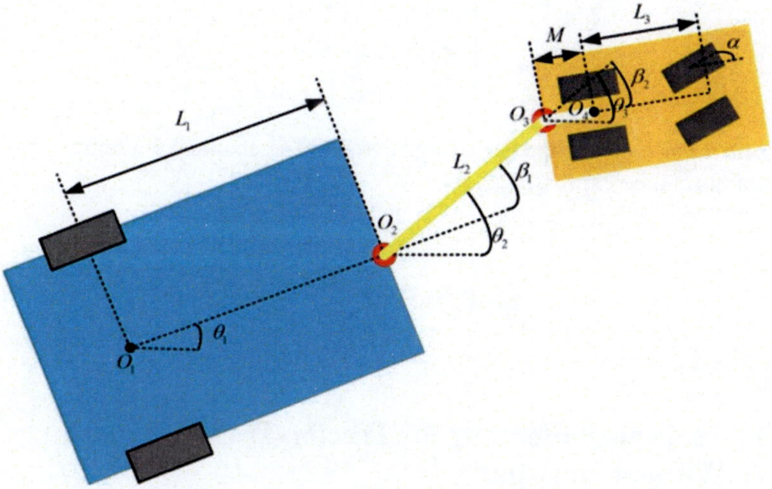


**Fig. 11.2** An off-axle hitching tractor-trailer system with drawbar

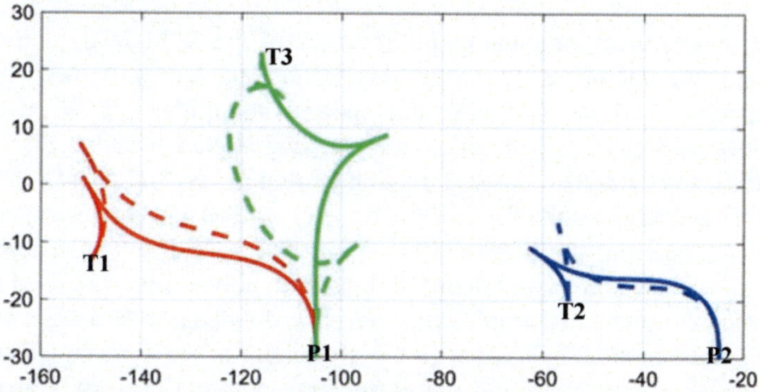


**Fig. 11.3** The trajectory of tractor-trailer system and tractor, the solid line represents the result obtained by SPM, and the dotted line represents the result obtained by PM

$$\frac{d}{dt}\begin{bmatrix} x_1 \\ y_1 \\ \theta_1 \\ \beta_1 \\ \beta_2 \end{bmatrix} = \begin{bmatrix} u_2 \cos\theta_1 \\ u_2 \sin\theta_1 \\ \frac{u_2 \tan\beta_1}{L_1} \\ u_2\left(\frac{L_3 \tan\beta_2 - M \tan\alpha}{L_2 \cos\beta_1 (L_3 + M \tan\alpha \tan\beta_2)} - \frac{\tan\beta_1}{L_1}\right) \\ u_2\left(\frac{L_2 \tan\alpha - L_3 \sin\beta_2 + M \tan\alpha \cos\beta_2}{L_2 \cos\beta_1 \cos\beta_2 (L_3 + M \tan\alpha \tan\beta_2)}\right) \end{bmatrix} \tag{11.5}$$

On the one hand, the taxiing velocity $v_1$ should be restrained within a proper range in case of emergency braking. On the other hand, $\beta_1$ and $\beta_2$ must be bounded to avoid an undesirable jackknife effect. Hence, the following state constraints should be considered:

$$\begin{cases} v_{1\min} \leq v_1 \leq v_{1\max} \\ \beta_{1\min} \leq \beta_1 \leq \beta_{1\max} \\ \beta_{2\min} \leq \beta_2 \leq \beta_{2\max} \end{cases} \tag{11.6}$$

Control variables should be restrained to avoid mechanical limitations. The control constraints can be expressed as

$$\begin{cases} u_{1\min} \leq u_1 \leq u_{1\max} \\ u_{2\min} \leq u_2 \leq u_{2\max} \end{cases} \tag{11.7}$$

## 11.3 Trajectory Planning for Tractor-Trailer System Without Drawbar

### *11.3.1 Experimental Environment and Parameters*

In order to verify the efficiency and accuracy of the SPM in trajectory planning problem for the tractor-trailer system, this section analyzes the experimental results with a specific example. The maximum speed of the trailer is 1 m/s, the maximum acceleration is 3 m/s$^2$, and the maximum steering angle ($\beta_1$ and $\beta_2$) is 0.9 rad. The tractor-trailer system is an on-axle traction system, $L_1$ is 7 m, and $L_2$ is 2 m.

The "time-energy optimal" objective function will be used in this experiment. The convergence index of the SPM is $10^{-4}$, $\mathbf{R} = \text{diag}(1, 1)$.The whole time interval is divided into 20 segments, and then 5-th Legendre polynomials are used for interpolation in each segment, the whole time interval is divided into 100 segments with 101 discrete points. In order to compare the performance of the two methods more fairly, the parameters of the SPM and pseudospectral method (PM), such as the number of interval segments, the interpolation order of each interval, and the convergence index, are set to be the same.

### *11.3.2 Experimental Results*

In the experiment, 3 tractor-trailer systems are selected for research. The tractor-trailer system 1 and 3 are towed from the same starting point (P1) to the predetermined position (T1 and T3), and the tractor-trailer system 2 is towed from the other starting point (P2) to the predetermined position (T2). Since the focus of this experiment is on the state of the trailer system, the terminal angle of the tractor can be set as free, and the other four state variables are fixed terminal states. The details of the initial and terminal states are set as follows:

$$\text{P1 - T1: } \mathbf{X}(t_0)=\left(-105\text{m},\ 30\text{m},\ 90°,\ 0°,\ 0°\right)^{\mathrm{T}}\ \mathbf{X}(t_f)=\left(-150\text{m},\ -13\text{m},\ 45°,\ \text{free},\ 0°\right)^{\mathrm{T}}$$
$$\text{P1 - T3: } \mathbf{X}(t_0)=\left(-105\text{m},\ 30\text{m},\ 90°,\ 0°,\ 0°\right)^{\mathrm{T}}\ \mathbf{X}(t_f)=\left(-116\text{m},\ 23\text{m},\ -90°,\ \text{free},\ 0°\right)^{\mathrm{T}}$$
$$\text{P2 - T2: } \mathbf{X}(t_0)=\left(-25\text{m},\ 30\text{m},\ 90°,\ 0°,\ 0°\right)^{\mathrm{T}}\ \mathbf{X}(t_f)=\left(-55\text{m},\ -20\text{m},\ 90°,\ \text{free},\ 0°\right)^{\mathrm{T}}$$

In order to balance the relationship between the terminal time (Mayer term) and the energy (Lagrange term), $k = 0.1$ in this experiment. According to the SPM proposed in this chapter and PM, the trajectories of 3 tractor-trailer system can be shown in Fig. 11.3.

In the above figure, the yellow line represents the trajectory obtained by the SPM proposed in this chapter, and the red line represents the trajectory obtained by the PM. It can be analyzed from the above figure that the trajectories obtained by the two methods meet the boundary constraints, which shows that the trajectories obtained by the optimal control method can meet the terminal constraints. In addition, the trajectories of the system 1 and 2 are relatively smooth, but the process of the system 3 and the corresponding tractor is relatively disordered at the terminal stage by PM.

To further analyze the effectiveness of the proposed method, this chapter gives the control variables of the tractor-trailer, the details can be shown in Fig. 11.4.

Figure 11.4a–c are the control variables of the traction system from P1 to T1, T3 and P2 to T2, respectively. It can be analyzed from the above figure that the change of control variables corresponding to the optimal trajectory obtained by the SPM in this chapter is relatively smooth, $u_1$ are located in $[-0.8, 0.8]$, and the steering angle is located in $[-38.66°, 38.66°]$, which means that there is no sharp turning in the whole process of movement. The acceleration $u_2$ is located in $\left[-0.5\,\text{m/s}^2, 0.5\,\text{m/s}^2\right]$, the deceleration and acceleration process is relatively smooth, and the acceleration in most processes is $0\,\text{m/s}^2$, and there is no sudden braking in the whole stage. The above results can prove that the control corresponding to the optimal trajectory meets the requirements, it's relatively stable and meets the engineering practice. However, the control variables, obtained by PM, corresponding to the trajectory from P1 to T3 oscillate seriously at the terminal stage, which is beyond the constraint, so it can be regarded as an unfeasible solution.

To further analyze the movement of the tractor-trailer system in the traction system, the speed and steering angle of the trailer can be obtained according to the state variables, it can be shown as the Fig. 11.5.

Figure 11.5a–c are the speed and steering angle of the trailer from P1 to T1, T3 and P2 to T2, respectively. Combined with Fig. 11.4 and 11.5, it can be analyzed that the trailer keeps constant motion with the maximum speed of 1 m/s or the minimum speed of -1 m/s in most stage, the speed is 0 m/s at the terminal point, and there is no sudden brake in the motion process along the optimal trajectory, which reduces the accident risk. When the PM is used, the trailer speed and steering angle from P1 to T1 and from P2 to T2 is relatively smooth, but the steering angle from P1 to T3 is obviously beyond the constraint (Fig. 11.3 and Table 11.1).

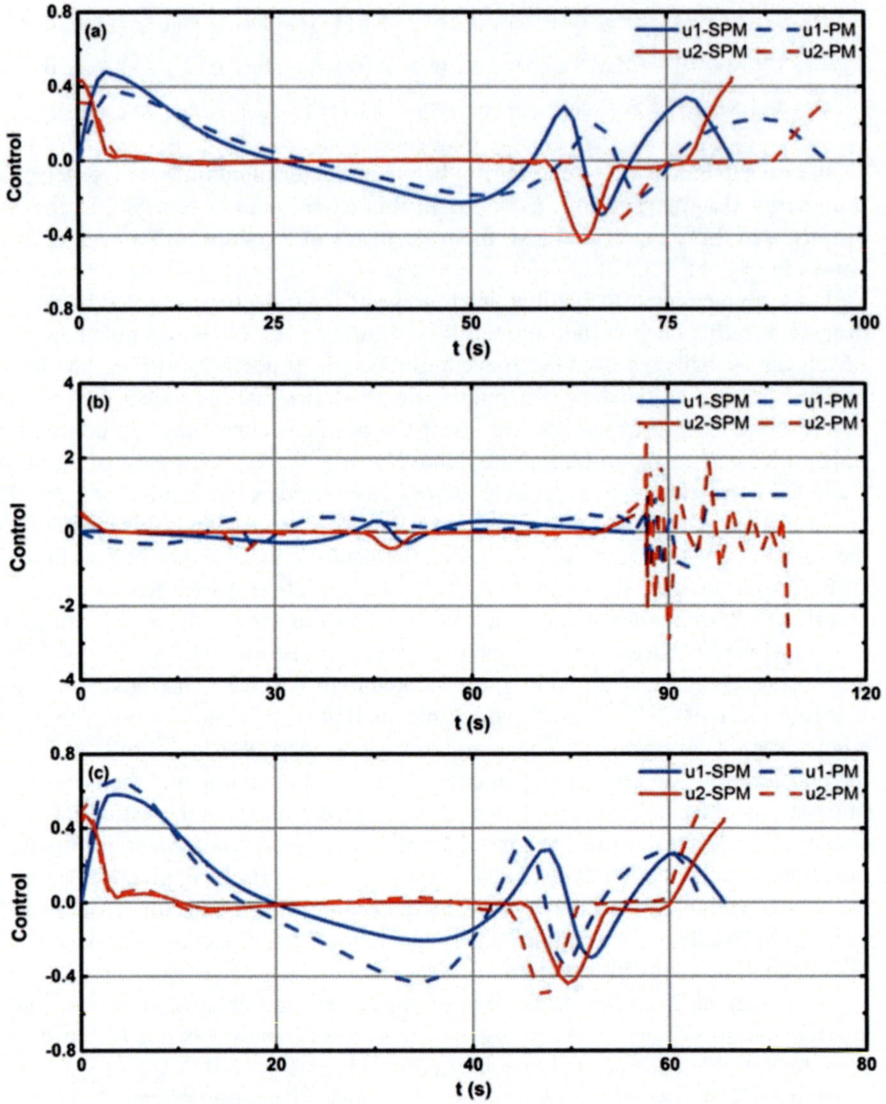


**Fig. 11.4** The control of the system

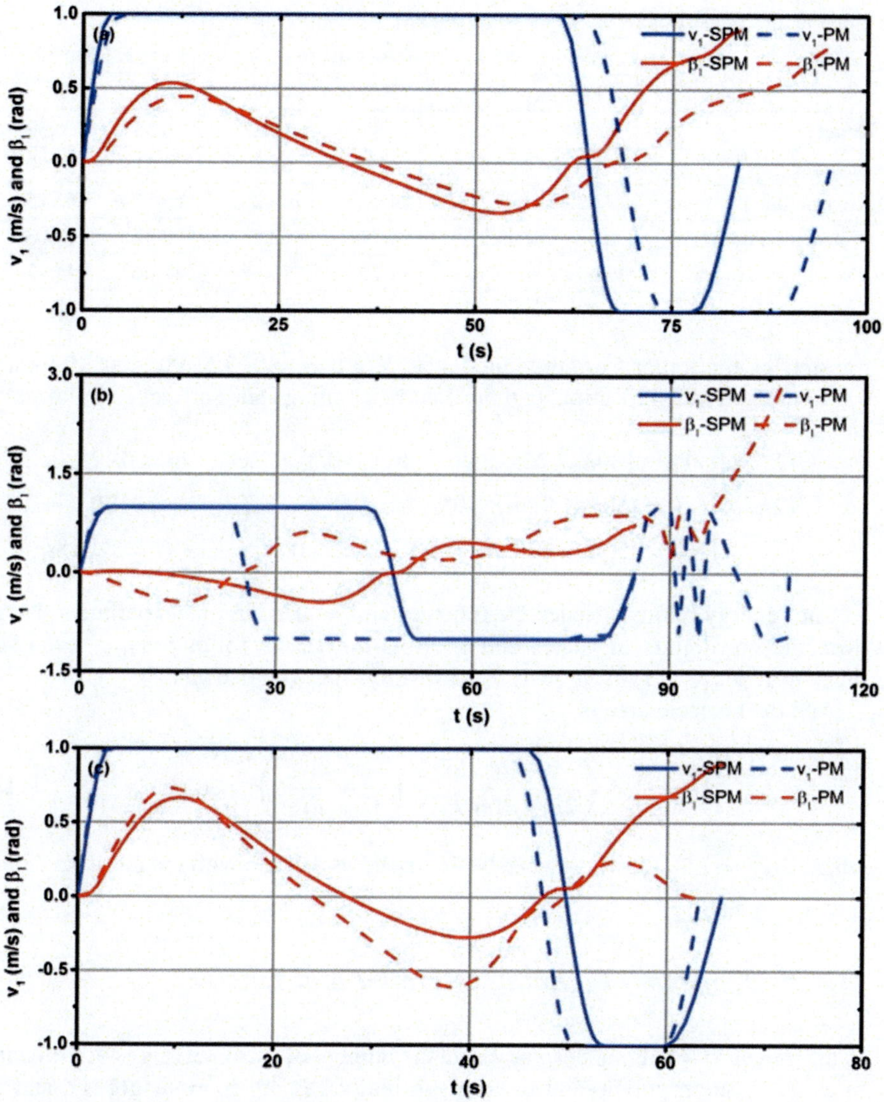


**Fig. 11.5** The velocity and steering angle of the trailer

## 11.4 Trajectory Planning for Tractor-Trailer System with Drawbar

### *11.4.1 Parameters Setting*

According to the kinematic model of the tractor-trailer system with drawbar, the system parameters are set as $L_1 = 6\,\text{m}$, $L_2 = 4\,\text{m}$, $L_3 = 2\text{m}$, $M = 1\,\text{m}$. We aim at

**Table 11.1** Comparison results of the SPM and PM

| | Case 1: P1 to T1 | | Case2: P1 to T3 | | Case 3: P2 to T2 | |
|---|---|---|---|---|---|---|
| | SPM | PM | SPM | PM | SPM | PM |
| Mayer | 8.33 | 9.53 | 8.47 | 10.83 | 6.55 | 6.32 |
| Lagrange | 5.30 | 3.81 | 4.24 | 65.54 | 5.51 | 7.80 |
| Arrival time (s) | 83.29 | 95.27 | 84.69 | 108.26 | 65.55 | 63.25 |
| Calculation time (s) | 5.33 | 45.27 | 4.78 | 68.63 | 2.32 | 21.03 |
| Performance index | 13.63 | 13.34 | 12.71 | 76.37 | 12.06 | 14.12 |

transferring the trailer from the initial state $\mathbf{X}(t_0) = (50, 7.5, 90°, 90°, 0)^{\mathrm{T}}(t_0 = 0\,\mathrm{s})$ to the 3 terminal location, and the 3 corresponding states are set as follows:

$$\begin{aligned}
&\mathrm{T1:}\ \mathbf{X}(t_f) = (130\mathrm{m},\ 26\mathrm{m},\ 90°,\ \mathrm{free},\ 0°)^{\mathrm{T}} && t_f = 142.51\mathrm{s}\\
&\mathrm{T2:}\ \mathbf{X}(t_f) = (50\mathrm{m},\ 98\mathrm{m},\ 90°,\ \mathrm{free},\ 0°)^{\mathrm{T}} && t_f = 120.44\mathrm{s}\\
&\mathrm{T3:}\ \mathbf{X}(t_f) = (55\mathrm{m},\ 62.5\mathrm{m},\ -90°,\ \mathrm{free},\ 0°)^{\mathrm{T}} && t_f = 84.18\mathrm{s}
\end{aligned}$$

The "energy optimal" objective function will be used in this experiment. Moreover, the constraints on states and controls are set as follows, $v_{1\max} = 1\,\mathrm{m/s}$, $\beta_{1\max} = 1$, $\beta_{2\max} = 1$, $u_{1\max} = 1$, $u_{2\max} = 1\,\mathrm{m/s}$, $v_{3\max} = 1.6\,\mathrm{m/s}$.

And the obstacle area is

$$1 - \left(\frac{x - 74}{23 + dist}\right)^4 - \left(\frac{y - 10}{17 + dist}\right)^4 \le 0 \tag{11.8}$$

where $dist = 1\,\mathrm{m}$, the system can be described as a circle with the radius is 13 m.

## 11.4.2 Trajectory Planning of Trailer and Tractor

In the trajectory-planning design, we set the number of sub-intervals as 5; the degree of approximation polynomial in each sub-interval is 40, $\mathbf{R} = \mathrm{diag}(1, 1)$, and the acceptable tolerance is $\varepsilon = 10^{-4}$. The trajectory of the trailer and real tractor can be obtained by using SPM, and profiles of the trailer's trajectory can be given in Fig. 11.6.

It can be analyzed from the above figure that the trailer 1, 2, and 3 are pulled out by the tractor firstly, and then it's pushed into the terminal position by the tractor. The trajectories are very smooth, and they meet the constraints of initial and terminal position. And the control can be obtained as follows:

Figure 11.7 a–c represent the control of the system along the optimal trajectory-1, trajectory-2, and trajectory-3, respectively. It can be analyzed from the above figure that the control variables corresponding to the three optimal trajectories are

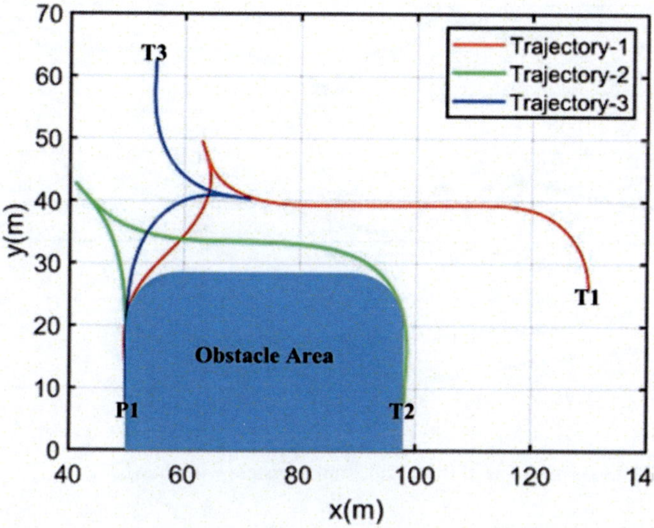


**Fig. 11.6** The trajectory of the trailer for the tractor-trailer system with drawbar

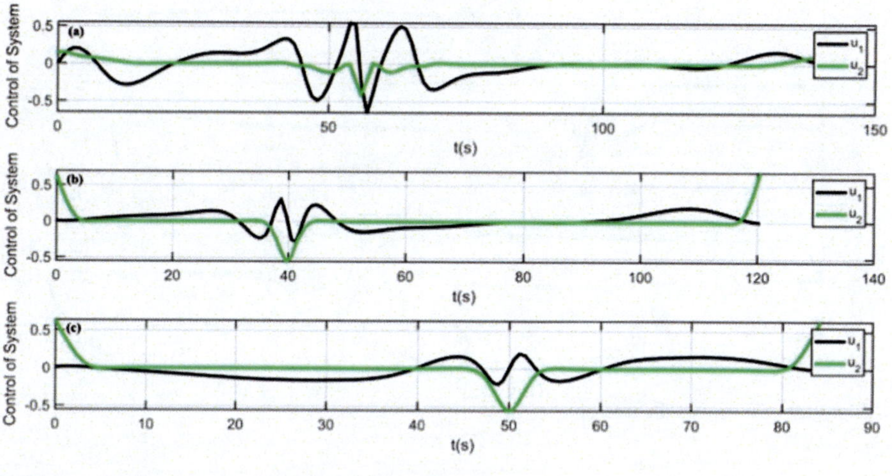


**Fig. 11.7** Control of tractor-trailer system with drawbar

relatively smooth, and they meet the corresponding control constraints. To test the fact that whether the results satisfy the state constraints, the steering angle is obtained as follows:

Figure 11.8a–c represent the steering angle of the system along the optimal trajectory-1, trajectory-2, and trajectory-3, respectively. During the process of movement along the optimal path, the heading angle is relatively smooth, there are no sharp turns, which reduces the risk of accidents caused by sharp turns. And the steering angle is less than 0.6, which falls in $[-45°, 45°]$, and it meets the corresponding state constraints.

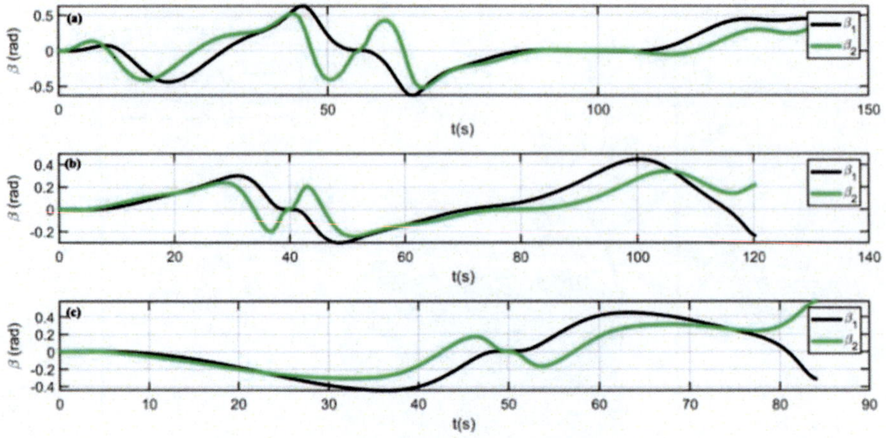


**Fig. 11.8** The steering angle of the tractor-trailer system with drawbar

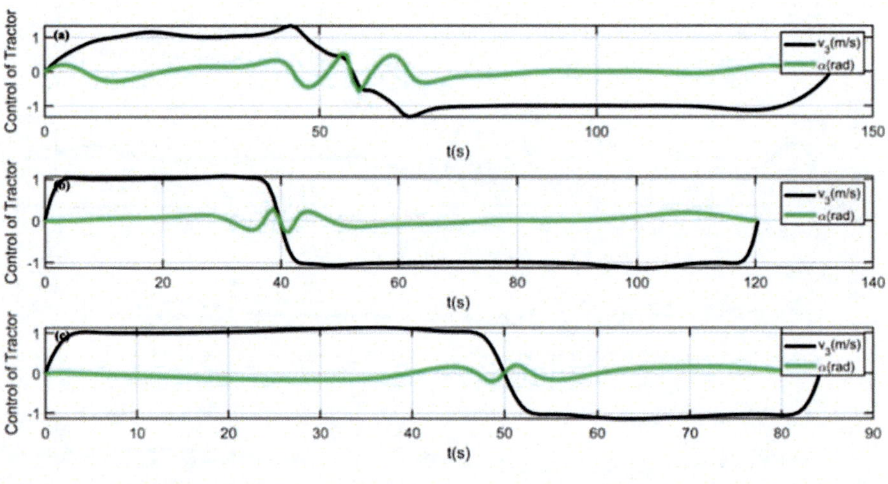


**Fig. 11.9** Control of the tractor

Figure 11.9a–c denotes the control of tractor along the optimal trajectory-1, trajectory-2, and trajectory-3, respectively. The velocities of tractor fall in $[-1.6\,\text{m/s}, 1.6\,\text{m/s}]$, which meets the constraint of velocity. There is no mutation of velocity, which reduces the risk caused by the drastic change of velocity. And the steering angle of tractor located in $[-45°, 45°]$, which meets the constraint of steering angle too.

The results show that both the trajectory and corresponding state variables are continuous and relatively smooth, and constraints can be satisfied by using the proposed SPM.

## 11.5 Conclusion

In this chapter, the trajectory planning problem of the tractor-trailer system is modeled, and the SPM is used to solve the model. The experimental results show that the proposed SPM can solve the optimal control problem of tractor-trailer system trajectory planning with high accuracy and efficiency, and the optimal terminal time can be obtained, the corresponding trajectories are also very smooth, and the control variables are in accordance with the engineering permission. Moreover, the SPM requires low initial values of covariates and control variables, and it has strong operability and applicability.

The application of the SPM in trajectory planning for the tractor-trailer system has many advantages, the details are as follows:

(1) The optimal control model of trajectory planning is established, and the optimal control method is used for trajectory planning. Compared with other methods, the results of the SPM can not only meet the terminal constraints, but also meet the optimal control conditions, so as to ensure the optimal results.
(2) The SPM can solve the optimal control problem with uncertain time with high accuracy, high calculation efficiency and fast convergence.
(3) Most of the traditional methods only consider the shortest trajectory, but without considering the control problem, it's likely to be unable to design the corresponding control which can meet the relevant constraints, so the corresponding results can't be applied to the actual project. Therefore, it's difficult to guarantee that the shortest distance trajectory is equivalent to the shortest time trajectory in practical problems. However, the model established in this chapter and SPM can describe and solve this problem effectively.

What's more, the different initial solutions of the state variables in the SPM will affect the convergence speed. How to solve this problem is the research direction of the follow-up work, and the algorithm can also be optimized to further improve the calculation efficiency.

## References

1. Si W, Qi Y, Han W (2015) Shipboard tractor-trailer system hangar Dispatching Planning Based on convex hull algorithm integrated with Dijkstra. Syst Eng Electron Technol 37(3):583–588
2. Han W, Si W, Ding D et al (2013) Multi trajectory dynamic planning of shipboard tractor-trailer system based on clustering PSO algorithm. J Beijing Univ Aeronaut Astronaut 5:610–614
3. Wu Y, Qu XJ (2013) Trajectory planning for taxi of tractor-trailer system launching. Science Chine: Technol Sci 56(6):1561–1570
4. Wu Y, Qu X (2015) Obstacle avoidance and trajectory planning for tractor-trailer system launching. Chin J Aeronaut 28(3):695–703
5. Wu Y, Hu N, Qu X (2019) A general trajectory optimization method for tractor-trailer system taxiing on flight deck of carrier. Proc Inst Mech Eng Part G: J Aerosp Eng 233(4):1340–1353
6. Johnston JS, Swenson ED (2010) Feasibility study of global-positioning-system-based tractor-trailer system- flight-deck persistent monitoring system. J Tractor-trailer Syst 47(5):1624–1635

7. Karkee M, Steward BL (2010) Study of the open and closed loop characteristics of a tractor and a single axle towed implement system. J Terrramech 47(6):379–393
8. Yuan J, Sun F, Huang Y (2015) Trajectory generation and tracking control for double-steering tractor-trailer mobile robots with on-axle hitching. IEEE Trans Industr Electron 62(12):7665–7677
9. Yuan J (2017) hierarchical motion planning for multi-steering tractor-trailer mobile robots with on-axle hitching. IEEE/ASME Trans Mechatron (4):1
10. Yuan J, Yang S, Cai J (2018) Consistent path planning for on-axle-hitching multi-steering trailer systems. IEEE Trans Indust Electron 1
11. Liu Z, Yue M, Guo L et al (2019) Trajectory planning and robust tracking control for a class of active articulated tractor-trailer vehicle with on-axle structure. Europ J Control
12. Ljungqvist O, Evestedt N, Axehill D, et al (2019) A path planning and path-following control framework for a general 2-trailer with a car-like tractor. J Field Robot 36(8)